Bernadette Vallely
Bernadette is an environmental writer, campaigner and broadcaster.

She has received many awards for her community work, ecological campaigning and writing, over more than thirty-five years. Bernadette was named a United Nations Global 500 Environmental Laureate in recognition for her outstanding support of global environmental action.

Amy Charuy-Hughes
Amy is a fashion graduate and eco-advocate. She works as an environmental fashion consultant and eco-designer of 'Greener Together with Amy'. Amy volunteers at Refill Wandsworth, a non-profit organisation working to promote free on-the-go drinking tap water in the UK.

Bethan Stewart James
Bethan is an inspirational nine-year-old Welsh-speaking eco-activist.

A member of her school eco-group, she has promoted her Green Schools Agenda in her hometown in Pembrokeshire, Wales.

Follow the authors' journey: www.yourplanetneedsyou.org.uk

Use #yourplanetneedsyou on social media

BERNADETTE VALLELY
WITH AMY CHARUY-HUGHES AND BETHAN STEWART JAMES

YOUR PLANET NEEDS YOU!

AN EVERYDAY GUIDE TO SAVING THE EARTH

virago

This book provides advice on how to lead a more respectful life towards the planet. It is not a substitute for medical advice from your healthcare providers. If you have any allergies, intolerances or medical conditions please consult a health practitioner.

VIRAGO

First published in Great Britain in 2020 by Virago Press
This paperback edition published in 2021 by Virago Press

1 3 5 7 9 10 8 6 4 2

A CIP catalogue record for this book is available from the British Library.

ISBN 978-0-349-01389-3

Typeset in Palatino by M Rules
Printed and bound in Great Britain by Clays Ltd, Elcograf S.p.A.

Papers used by Virago are from well-managed forests and other responsible sources.

Virago Press
An imprint of
Little, Brown Book Group
Carmelite House
50 Victoria Embankment
London EC4Y 0DZ

An Hachette UK Company
www.hachette.co.uk

www.virago.co.uk

To Michael, my beloved son, who needs this book, as we all do. Bernadette

To my niece River, age 4, River's friends and her generation. Amy

Thank you, Greta Thunberg, Mum, Dad, Dylan and Flossie, Eliza and Leila. Bethan

IMAGINE A WORLD

Imagine a world that's clean all around,
People work together. Happiness is found.
Let's show respect, it's who we are.
Let's not treat our earth like a disposable jar.

When all the plants are healthy and green,
When all the seas are thriving and clean,
That is what our earth could be.
Why don't you come and help out me?

If you could see what I can see,
When the world is not a catastrophe,
Then you would 100% help me!

Bethan Stewart James, age 9, January 2020

PREFACE

If you're reading this book it means you care. It means you're probably worried about our world and you can see around you the effect our lifestyle is having on Planet Earth.

Thank you, because if there was ever a time to stand up for Mother Earth, this is the time. The planet is in trouble. We, as human beings, are in trouble. Our relationship with the earth is strained to breaking point.

This book is part of the solution. Here are nearly one hundred of the most pressing environmental problems that affect us today along with information that is geared towards solutions, alternatives and better outcomes. Most crucially, this book is designed to inspire action, commitment and community.

The coronavirus pandemic put us all on notice. It has humbled many of us. It has brought grief and sadness and affected every family across the world in some way, but it has also brought nature closer to our attention, and created a new respect and awe for our world in many people.

Overnight aeroplanes were grounded, as were non-essential car journeys. Within days, the air became cleaner and fresher. It was as clear as the sky that fossil fuels are the main cause of air pollution. The protests and lies of the industries that promote them evaporated overnight, their well-funded arguments in tatters. We should not accept seven million deaths from air pollution and the fossil fuel industry each year; these are needless deaths.

Our expectations and relationship with the natural world may never be the same again.

When people have emerged from the pandemic stay-at-home orders across the world they have often described deep feelings of love, respect and gratitude for their local parks, rivers and nature reserves and for nature in general. They realised how connected we all are.

Research has revealed the immediate link between our destruction of the ecosystems around us and our fragile health and food supply.

As human beings we find it hard to see the connections between our own personal actions, the environmental emergency and our own lives and ultimately deaths.

Our lives have changed for ever because of the fragility and vulnerability of our species, and our arrogant and false belief that we can choose how we engage with nature.

This often wilful ignorance, unwillingness and inability to act together means an immense and serious environmental emergency has been building for years. We pretended it wasn't as bad as the statistics said it was. We learned to live with it.

It has become very clear that reducing polluting emissions from transport and industry dramatically changes the day-to-day living conditions and air quality for millions of people. During lockdown measures we could see cities without smog. For the first time in thirty years people could see the Himalaya Mountains over Kathmandu. The skies above Los Angeles were clear in less than two weeks.

Our choices after the pandemic emergency subsides – choices of transport for work, reasons for air travel, how we shop, how factories operate – will determine whether this window into a healthier urban future was a footnote or a foreword.

We could see how animals and creatures all around us reacted. They ventured back into urban areas and streets, and ate our flowers, the grass, privet hedges and anything they could find. Animals enjoyed having a rest from the constant noise and intrusion of human beings and their machines.

People who survive a serious accident such as a car crash re-evaluate everything in their lives – not just the causes of the accident, but what they value, which relationships they should have given more time to, which didn't matter and the choices they've made over their life.

Coronavirus is humankind's car crash. Sometimes it takes a

massive jolt for people to wake up, to make changes and take action.

Everything that you do affects the planet.

It's time for you and me to take some positive steps together so that we can be part of the solution, not part of the problem. Everything you do can be positive, ethical and supportive to the earth. It is our home. There is no alternative, no Planet B.

So thanks for reading this book – we hope it inspires you to do something, to start somewhere. No matter how small an action you take, when you open your mind and decide to become part of the solution you are on a journey that might never end. We hope to see you along that path with all the others who are working towards a greener, fairer, more sustainable future.

Bernadette, Amy & Bethan, May 2020

CONTENTS

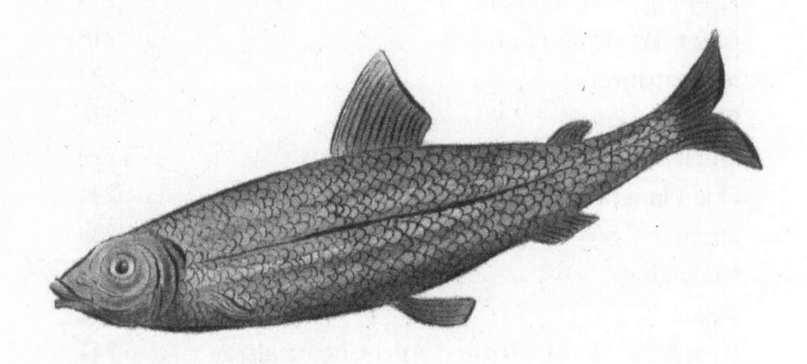

QUIZ: HOW ENVIRONMENTALLY AWARE ARE YOU?

Here is a quiz to see how well informed you are about the environment and maybe make you think a little. It's a bit of fun, not too serious. Try it now, again when you have finished reading the book, and again in six months' time, when you have taken some action and learned more as you develop your green understanding.

1. Everybody on the earth has the capacity to be an earth steward, an earth protector. What do you think are the most important values of an earth steward today?
   a. Telling the truth, using science and reason to win an argument and change someone's mind.
   b. Touching the earth, getting a connection with the earth, an understanding.
   c. Fighting to protect your local nature area – the park, river, woodland or wildlife area near you.

2. In which area of business would you be able to save energy, increase recycling and lower emissions with the highest impact on the planet?

a. Building renovations and housing upgrades.
b. Old landfill sites that have no recycling or sorting.
c. Concrete waste from building our motorways and bridges.

3. In its simplest form, we are all connected. If a butterfly can flap its wings on one side of the world and have the potential to change the weather and cause a storm across the oceans in another part of the world, then what can you do that matters MOST?
   a. We can't do anything alone – it is up to the politicians to change the laws, then I will be forced to reduce my carbon pollution like everyone else.
   b. We can only do a bit; but everyone has to be on board before we do our bit too, otherwise it's not fair.
   c. Reducing my carbon output means making changes in my life; I am taking steps towards that now to lead the way.

4. Which statement matches your shopping experience the most? My personal shopping habit is . . .
   a. Based around fulfilling an unconscious need.
   b. An absent-minded filling of time.
   c. Quick as possible and loathing it.
   d. I never ever shop.
   e. A pleasure – I know all my suppliers and enjoy my relationships with them.
   f. Craving only one thing week to week, and repeat buying that one thing.
   g. Only online for every single thing – I shop when I want and immediately.

5. Waste and Recycling is a big topic. In your household what would you say is the biggest waste problem?
    a. Too much food being bought and thrown away.
    b. Too many clothes and fashion items cluttering up the cupboards.
    c. Obsessive hoarding of one type of item: machines, crockery, clothes, books, magazines, etc.
    d. A huge collection of expensive lotions and cosmetics, perfumes and creams.
    e. Too much packaging and electronic waste from online purchases, including mixed-waste boxes, cables and cardboard.
    f. Animal waste: animal items, poo bags and clumps of cat litter.

6. Switching off electrical appliances is a good thing because:
    a. Your body can rest from the constant pulse of electricity and find instead the pulse of nature and the earth.
    b. You can save lots of money, more if you put effort into doing it.
    c. You get into the habit of not wasting resources, and not creating carbon pollution when you don't need to.

7. Transport is vital for our lives and progression. Whether it's getting to work, school, even visiting family, you need to work out how to do it. What are your priorities when you travel?
    a. I like to take a train and watch the world go by with a cup of tea.
    b. I like to cycle or walk and be in control of every part of my journey and at zero cost.

c. I car share to work; it's friendly and saves petrol and money.
d. I love the bus and it stops near my home; it's the cheapest form of travel for me.

8. What is the most common form of litter that ends up in the oceans?
   a. Cigarettes and cigarette ends.
   b. Red caps from cola bottles and tiny bits of plastic.
   c. Broken flip-flops, beach balls and sandcastle-type toys.
   d. Tissues, paper and newspaper.

9. Which of the following creatures are at risk of extinction or endangered in Britain?
   a. The Gwyniad, a freshwater whitefish native to northern Wales.
   b. The Song Thrush, found in forests, gardens and parks across Europe.
   c. The Shining Ram's Horn Snail, a freshwater snail found in ditches in Britain.
   d. The Shrill Carder-bee, native to Britain and found across Europe.
   e. The Southern Damselfly, a European fly found in rivers and freshwater springs.

QUIZ RESULTS: ADD UP YOUR SCORES!

1. All three answers are technically correct; your choice reveals more about your path to protecting the earth and what kind of actions would interest you.

Answer A: Score 2 – You are able to find like-minded people in scientifically-based organisations and communities.
Answer B: Score 2 – You are a gardener of some kind. Getting your hands into the earth, growing something is key to your happiness.
Answer C: Score 2 – You are likely to be a regular walker – maybe you have a dog? You might pick up litter as you walk? You notice your local area and have respect for wildlife.

Additional scores:
Score 1 if you are a member of any charity in your local area.
Score 3 if you are an active member of an environmental charity or group.
Score 0 if none of the answers apply to you.

2. Answer A: Score 2 – Refitting, renovating and using sustainable building materials in housing would not only make a difference overall to the planet but would also increase the availability of housing.
   Answer B: Score 3 – In the future the tens of thousands of landfill sites in the UK will need to be opened and mined for their raw materials, which will include many mixed metals, rare-earth materials and other valuable items.
   Answer C: Score 1 – Industry overall produces the most waste but concrete recycling into aggregate or chipping for reuse in the building industry is already well established.
   Score 0 if you did not understand the question or answers.

3. Your answer will show whether you value your independence or whether you prefer being part of a group. Are there any parts of your life where you can take the lead on the environment today?

Answer A: Score 1 – You are about five years behind the curve as laws are usually very slow at creating action and are a last resort. The actions you are taking today are becoming less relevant as the world is already moving towards a greener economy. This book should give you hundreds of ideas on how to start if you want to join in.
Answer B: Score 2 – If you think you can only do a 'bit' then you will stop yourself from taking proper green actions in your life but at least you are doing something. This book should give you more ideas for action in your community.
Answer C: Score 3 – You don't mind leading the way on improving the environment. Reducing your carbon output means making changes in your life; you are taking steps towards that now to lead the way.
Score 0 if you didn't understand the question yet. Reading this book will help you understand many of these ethical and ecological issues.

4. Most people have a practical and emotional attitude to shopping. Notice how it feels in this exposure of your habits.

   Answer A, B or G: Score 0 – There are expressions of impatience and unconscious shopping in this group.
   Answer C or F: Score 1 – A quick shopper and unlikely at least to buy extra or unnecessary items.
   Answer E: Score 2 – Buying from small-scale and family businesses means you have a direct relationship with your supplier.
   Answer D: Score 3 – It is an unusual and lucky person indeed who is able to shun all forms of commerce.

5. Tackling long-term waste issues in your household can be complex. Start somewhere.

Score 3 if no one in your house has a problem.
Answer A: Score 0 – 1.9 million metric tonnes of food is wasted in the UK each year.
Answer B: Score 0 – Billions of pounds' worth of clothing lies in drawers, unopened.
Answer C: Score 1 – Hoarding and obsessional collecting can become a problem. Share your wealth.
Answer D: Score 0 – British women spend £10 billion on cosmetics every year.
Answer E: Score 1 – Packaging waste is growing. Take electronic items to a repair centre.
Answer F: Score 1 – If you use an environmentally-friendly, animal-waste version of litter.
Score 2 if your pet poops outdoors and you pick it up in a compostable bag or with a newspaper.
Score 0 otherwise: indoor animals contribute millions of tonnes of non-recyclable waste each year.

6. Answer: All the answers are correct. Put them in order of importance and you can see which areas are more important for you.

   Answer A: Score 3 – It's good to bring nature into your home now and then.
   Answer B: Score 1 – It is worth saving money.
   Answer C: Score 3 – The carbon reduction proposal saves money and the planet.
   Score 0 if you have been using electricity without thinking. There are loads of ideas in the book we hope will help you to understand what it's all about and witness the benefits of a greener perspective.

7. All these transport methods are acceptable for greener transport options. Walking always tops the list as the best green option.

   Answer A: Score 2 – For longer journeys there is no better way to travel.
   Answer B: Score 3 – Your own energy always number one!
   Answer C: Score 1 – If you have to use a car, then at least share it! Score 2 if your car is an electric vehicle.
   Answer D: Score 2 – The most popular form of short-distance travel, and a blessing for the majority of humanity who use a bus to get to work, school and home.
   Score 0 if you only use a private car without other passengers.
   Score 0 if you have taken an aeroplane for work in the past three years.

8. All of the items listed have been found on beaches and in the oceans.

   Answer A: Score 3 – The most toxic item and they include the plastic tips.
   Answer B: Score 3 – Birds eat the red tops thinking it is food: they die with this plastic inside them.
   Answer C: Score 2 – All this plastic ends up in the Great Pacific Garbage Patch eventually, or is broken down and eaten by sea-life.
   Answer D: Score 2 – Well, at least it is degradable! Please don't litter the beaches.

   Additional scores:
   Score 0 if you can remember leaving litter on a beach yourself, or throwing waste overboard on a boat.

Score 3 if you have ever picked up litter on a beach somewhere and put it in the bin.

9. All answers are correct: Score 2. All these creatures are at risk of extinction or in some way threatened in their habitats.
   Score 0 if you randomly chose an answer.

   Now add up your answers!

How Green Are You?

Mostly 0s – Palest Green: You are at the very beginning of your green awareness. You do not really have any idea what it all means!

Admit it, you are only reading this book because your daughter or son or aunt or mum, or someone who loves you, bought it for you, and you have no idea what it's all about and no interest in being green, yet … In fact, you're probably bored with all the fuss. Well, at least you're being honest, and you can start here. It's a good place to begin! This book has hundreds of ideas and plenty of information to help you. There is something for everyone.

Mostly 1s – Nearly Green: You are clear about some of the answers but confused about others.

If this description suits you: at least you have begun your own green journey. You might have done one or two things like recycling because the local council made you. Perhaps you are a keen cyclist but never realised the environmental brownie points you get by cycling? You are already part of the solution.

Keep up the good work because the planet needs you to keep improving!

Mostly 2s – Active Green: You know what it is all about.

If you are here, you are well along the green road – congratulations! An active environmentalist like you is achieving a lot. When you are open to being part of the solution you know there is a lot you can do personally to make the planet better.

You can improve, however: there are, of course, more lifestyle changes you can make to supercharge your life with green activism. We hope this book will show you how you can improve, perhaps in areas you have yet to think about. Good luck and keep up the good work!

Scored mostly 3s – You are a Greta Green! This is your life already – you live it, you breathe it, 24/7.

Greta Thunberg is going seriously green. She is regarded as an inspiration because she has chosen to do what the planet needs first rather than what is comfortable for her. Her lifestyle is based on One Planet Living – that is, living within the planet's capacity – she is using one planet's worth of resources.

Greta has taken a vow not to fly, not to buy new things and not to eat meat. She is a vegan, eating a plant-based diet. By being careful about how she lives, Greta shows us that yes, sometimes it is difficult, but not impossible, to be as green as the planet needs you to be. She is trying to live a zero-carbon lifestyle by not creating pollution as she travels.

You don't have to do everything Greta does. That is not the point. The point is that your journey is a conscious one – that you are doing your best and utmost to do the right thing for the earth, for your future, for everyone's future.

INTRODUCTION

What does being green really mean?

The word green means different things to different people and cultures, especially when it relates to environmentalism and protecting the planet. Politicians, business leaders, environmentalists and the general public all have their own definitions. Ours is based on these principles:

LOVE

When you have love and respect for everything around you, you can see and care for what happens to the planet.

If you love your pet dog or cat, you have the capacity to love all animals on earth. If you love your local park or your woodland walk, you have the capacity to love all trees. If you love swimming in the river or walking on the beach, you will know that nature pulls you in and allows you a greater sense of being. When you love nature you find awe and respect in your heart.

Everything is interconnected. Every breath we take makes us part of this chaotic, complex entity that is Planet Earth. And every act has a consequence, even if we don't live to see it. When you respect and love the earth you notice this and you make extra sure that your actions will not hurt people or nature, now or in the future.

SUSTAINABILITY

Creating a life where people live within their means is part of being sustainable. When a raw resource, such as a crop or an earth mineral, is being harvested a long-term view is needed. The aim is to leave the water, air and soil, the animal, bird and insect populations, as healthy as they were beforehand. Then they can regenerate themselves easily and with healthy abundance.

A sustainable business is one where quality counts more than quantity. The ability to repair or reuse a product being more important than its price is sustainability in action. Conserving the raw resources and the energy used to create a product makes the business fit for generations to come. A sustainable business will continue for generations without destroying the land, water and air around it. Sustainable business enhances the environment and creates products and waste that are non-threatening to the future.

YOUR TIME

Everything you do matters. Everything has a price, and your time is the most important thing you have. The way you spend your time and money is very important; it can literally affect your future.

Value time spent in nature. Value time spent in activities with people you love and doing things that bring you joy and pleasure.

It sounds so simple. It is.

You can also spend some time looking after the planet. Bringing compassion into your active social world will bring you joy. From gardening to litter-picking, there are so many ways to touch the earth and be at peace with nature.

Reading and informing yourself, campaigning, demonstrating, activism, writing and commenting on social media and on links to companies and government: these are all ways in which you can use your time productively to support the planet and your environmental goals. Volunteer for a local group; if there isn't one you like, set one up yourself.

PERSONAL ACTION

It doesn't matter where you begin, what matters is that you do begin. It doesn't matter what you know now, what matters is that you ask questions, that you search for answers and that you enquire about the very planet that you live on and need for your survival.

You can take action on your own or in a group; you can take action with a friend, in your workplace, your school or your university. At the end of the day, however, *you* must make personal choices about what you're going to eat, how you're going to live your life, the products that you buy and the books and newspapers you read.

Personal action will empower you; it will allow you to grow and learn, to develop. Whether it's learning to ride a bike; learning to cook, so that you can decide what you eat; planting a tree or buying second-hand or vintage clothes rather than new, environmental action is good for the planet and good for you.

GROUP ACTION

Your personal actions are important and there are many people who will join you.

In fact, millions of ordinary people are part of environmental groups – campaigning, protecting, litter-picking, letter-writing, demonstrating, planting and taking positive action day by day. Their aim is to protect the planet. Join them. Get your friends involved. Do something together and you can help to change the world.

There are groups that have already been set up to protect and campaign for the planet, like XR (Extinction Rebellion), Friends of the Earth and Greenpeace. They campaign for more action; they protect our natural heritage and important sites – as do others such as the National Trust – and they plant trees for the future with organisations like TreeSisters or the Woodland Trust.

There are organisations that support specific animals in need of protection in their natural habitat, from Orangutans in Borneo to wild birds in Orkney. African forest elephants, sand cats and pygmy hippos are all threatened with extinction and need support, campaigning and help in many ways from groups and networks of people willing and interested in their survival. Thousands of British birds, insects, mammals and other animals are in peril or even at risk of extinction today.

Each town, city and country has a network of positive people who have decided to do something to help, to stop the destruction, to encourage activism. These are the green heroes of the future. This is the activism that will rewrite history. (See page 338 for a list of environmental groups.)

INFORMATION IS POWER

Once you look, you will see environmental news and information everywhere. If you have a dedicated environmental centre in your town you will be able to meet other like-minded people interested in your issues in one place. You can find out where meetings are held, when the next activity is on, what you can do to help.

The internet, of course, has millions of options and answers. Type in a question: How can I travel from Leeds to Luton without creating carbon pollution – CO_2? What is the cutest endangered animal? How many trees do I have to plant to balance out the effect of travelling from London to America in an aeroplane? Can I survive if I make my own clothes?

Find out about things. Read. Learn. Inform yourself and be empowered with that information.

KEY SPECIES IN DANGER
Illustrations by Stu McLellan

Common Bottlenose Dolphin

Latin Binomial name and Species: ***Tursiops truncatus* – *T. truncatus***

A dolphin you are most likely to have seen in a water or ocean park as they have been captured and exhibited for many decades. They are found across the world, including the seas of Britain but mostly in warm water.

The Common Bottlenose Dolphin is the long-beaked dolphin you will have seen in pictures and in shows. They live for about seventeen years in the ocean and up to fifty years in captivity. They are considered to be highly intelligent creatures and have communicated with humans for decades. Dolphins are an important indicator species for the oceans surrounding the UK.

Gwyniad

Latin Binomial name and Species: *Coregonus pennantii* – *C. pennantii*

A freshwater whitefish native only to northern Wales. It has been found to be critically endangered in Bala Lake (Llyn Tegid) in Wales: see International Union for Conservation of Nature (IUCN) – Red List of Threatened Species 3.1.

This freshwater fish is critically endangered because of a number of threats, especially poisoned water in the lake, and the Ruffe, a fish introduced by humans in the 1980s. The Ruffe eats the eggs of the Gwyniad. Various conservation programmes are urgently trying to transfer the eggs to another nearby lake, Llyn Arenig Fawr.

Shining Ram's Horn Snail

Latin Binomial name and Species: *Segmentina nitida* – *S. nitida*

A freshwater snail from the Planorbidae family called an aquatic gastropod mollusc.

This snail lives in water weeds, marshy watery land and ponds as well as ditches. It is a rare snail and is listed as an endangered species in Britain. It has a distinctive sinistral coiling like a spiral. It is only about 7mm in diameter with a glossy shell. Habitat loss has threatened its survival.

Shrill Carder-bee

Latin Binomial name and Species: *Bombus sylvarum* – *B. sylvarum*

A bumblebee known also as Knapweed Carder-bee with widespread distribution across Europe, from the Ural Mountains in the east to Britain and Ireland in the west. The Shrill Carder-bee is from the Apidae family.

Its declining numbers have been of great concern across Europe because of intensive agricultural practices and habitat loss. It is found only in a few small areas of southern England and south Wales. It is a small, pale-yellow bee, once common across Britain.

Song Thrush

Latin Binomial name and Species: ***Turdus philomelos* – *T. philomelos***

The Song Thrush is found across Europe and in gardens, parks, forests and woodlands across Britain. It is sometimes migratory to North Africa, southern Europe and the Middle East.

One of the Song Thrush's biggest dangers is from household cats who kill many in the garden. This could be avoided if cats wore bells, which would warn the Song Thrush. They have declined in numbers recently to worrying levels because of intensive farming practices, including the ripping up of hedgerows and the overuse of pesticides.

Southern Damselfly

Latin Binomial name and Species: ***Coenagrion mercuriale* – *C. mercuriale***

The Southern Damselfly is from the Coenagrionidae family. It is listed as near threatened by the IUCN 3.1.

The Southern Damselfly inhabits rivers and freshwater springs and is threatened by habitat loss in Britain. It is found across Europe as far east as Slovakia and as far west as the United Kingdom. This damselfly has very distinctive markings on the second segment of the abdomen, which resemble the astrological symbol for the planet Mercury. Sometimes it is called Mercury Bluet because of this.

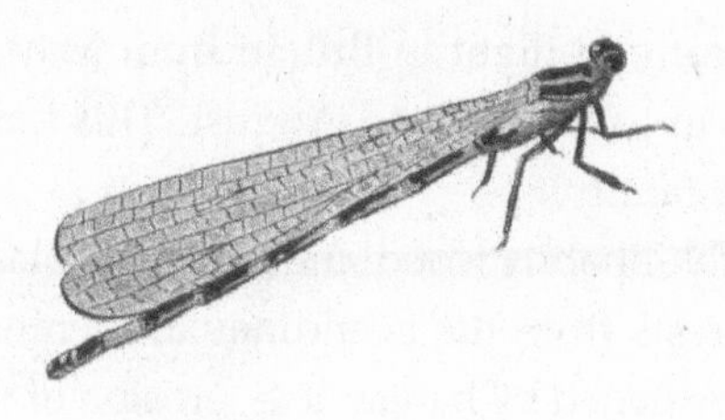

Stag Beetle

Latin Binomial name and Species: ***Lucanus cervus* – *L. cervus***

The stag beetle is listed as near threatened and is on the IUCN Red List. This beetle is from the Lucanidae family.

The stag beetle is found throughout Europe but not Ireland. It is found mainly and especially on hilly, sunny and mountainous areas. It was once common in mainland Britain.

Woodland Hoverfly

Latin Binomial name and Species: ***Chrysotoxum octomaculatum –C. octomaculatum***

A gentle hoverfly in flight in Britain from May to September with its peak in June, July and August. This hoverfly is from the Syrphidae family.

The hoverfly inhabits woodlands and scrublands as well as deciduous forests from the Scandinavian Peninsula to North Africa. It is threatened by habitat loss, erosion of the forests and increasing urbanisation of scrubland and other areas.

AN A–Z OF ENVIRONMENTAL INFORMATION

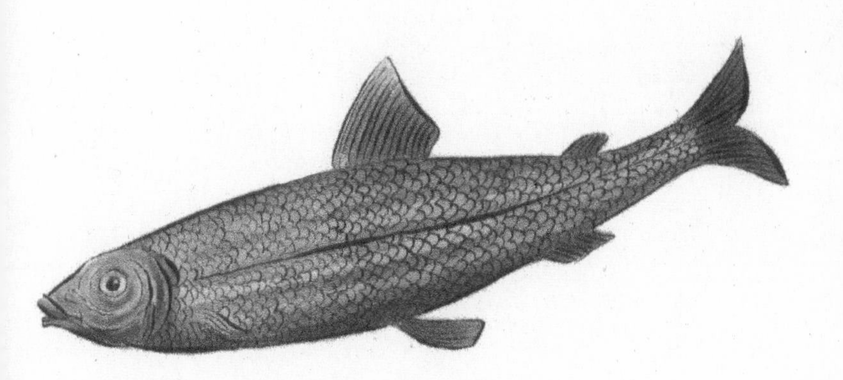

A

ACID RAIN

Acid rain is created when gases like sulphur dioxide, nitric oxide and nitrogen dioxide rise up into the atmosphere from exhaust fumes and from industrial processes like smelting steel and burning coal. The chemicals react with oxygen in the air and dissolve in water droplets, and when these fall down as rain or snow it affects everything it touches: trees, buildings, bodies of water and the soil. It literally burns what it touches with acid.

Thirty years ago, acid rain was a serious global problem. In Germany, for example, 50 per cent of the trees were affected. Fish and other aquatic life were dying in lakes throughout the world. The problems were scientifically recognised and environmental groups and government bodies started working together to find a solution. Countries had to pass Clean Air Acts. There were legal agreements to minimise the pollution and chemicals as much as possible so that the trees and lakes could survive. All industrialised nations had to address the problem.

It has been a long haul but lakes and forests are slowly starting to revert to their cleanest biological state, fish are returning in the affected rivers, and trees have been growing properly. This shows that when politicians, heads of industry and scientists work together they can pass laws and make changes that really do improve the environment.

Technology has changed and many countries are now burning coal using new methods that have reduced the pollution by more than 80 per cent. People came together to solve the problems and they made it happen.

Today, however, our oceans suck in carbon dioxide emissions from oil, gas and coal burning, and this makes them more acidic for marine life. Scientists have warned that the problem is putting the ecological balance of ocean life at risk.

What can you do?

Save Energy! Saving energy in every part of your life makes a difference to the world. Turn off the lights, switch off the television or your computer when they are not in use.

There are hundreds of ways you can use energy more efficiently. Cycle or walk, travelling under your own steam. Use public transport when you can. This will help cut down on car emissions.

Acid rain exists because chemicals are escaping into the atmosphere and water. We need better filters, more new inventions and cleaner processes. Business must invest today so all industries are clean.

SEE ALSO: Air Pollution, p. 29; Chemicals, p. 54; Transport, p. 294

ACTION, PROTEST AND EMPOWERMENT

Margaret Mead, a pioneering cultural anthropologist and scientist, said that we must never doubt the power of a few people to change the world. History only serves to confirm this.

To protest against injustice is an ancient right in Britain, and our history is full of examples of those who have exercised that right while being tortured or even killed. In 1834 the Tolpuddle Martyrs, farm labourers from Dorset, were shipped to Australia for starting a 'Friendly society', the first ever trade union of six men. Those who have fought for justice and equality in the workplace have since become the trades union movement.

Those who have fought for common health services and dignity in the community have become our NHS backed up by the charities that provide compassionate services in many walks of life and to people in the direst need. Those who cherished and loved animals have created thousands of organisations, shelters and projects to care for those creatures that have no voice.

At the beginning of the twentieth century women campaigned for a woman's right to vote in Britain, pushing for laws to allow them the same rights as men to vote and choose their political leaders. Many were imprisoned; some died. It took years of action but people joined together in street protests and in campaigns that changed our view of women in society. The seeds of these actions allowed modern feminism to develop.

We can see that now with hindsight. We can trace the development of women's action, which led to better conditions and rights later on. Sometimes such a development of consciousness takes place quickly; sometimes it takes years and many generations. It can be frustrating and inspiring all at once. In the case of women's rights, it has been a long and very slow path. The only place left in the world where women can't vote today is the Vatican City in Italy.

When an individual or a small group first stands up and speaks against what has become commonplace in society, they can be shouted down or laughed at. Those who see first are usually disbelieved. Forty or fifty years ago, the scientists and campaigners who spoke about climate change and species extinction were told they were just alarmists. They were ridiculed and sneered at. Yet everything they said has come to pass.

It is the members of non-governmental organisations and charities, pressure groups and action groups that push local opinion and political opinion to create social action.

Greenpeace, Oxfam, Friends of the Earth, WWF (World Wide Fund for Nature), XR (Extinction Rebellion), and PETA (People for the Ethical Treatment of Animals) are just a few of the thousands of organisations that exist in Britain dedicated to making the planet better. These groups are made up of individuals working together with shared goals and common aims.

When people get together and say there is a problem and we need to solve it, they are being positive, looking for a solution, willing to take action and to expend their own energy to solve that problem.

From animal protection to river cleaning, the number of organisations and individuals undertaking some form of environmental action is huge. And yet there is always space for more.

Fridays For Future, and all the associated school climate-strike movements that have sprung up in support of Greta Thunberg's work, are relatively new in the field of environmental protection and action. They are inspiring not least because very young people are involved; they have an energy and enthusiasm that is infectious. As Greta Thunberg says, 'This is our future.'

In some countries environmental action and protest will get activists arrested, perhaps even jailed – or worse. Environmental campaigners, protesters and activists, male and female of all ages, have been targeted by unscrupulous loggers, multinational companies, and even local people who have decided that money is more important than the natural environment in which they live. We know about just some of the environmental heroes that have gone missing, presumed dead or murdered, in their own countries for standing up in defence of Planet Earth. In 2017 some 197 individual environmental activists across the world were murdered for their beliefs.

In Britain we have laws that protect us and allow us the freedom to demonstrate. Many thousands of XR protesters were purposely arrested during their campaigns in 2019. They wanted specifically to push legal boundaries. They wanted to speak of their worries and fears of the environmental catastrophe that Planet Earth is now experiencing. They wanted to be arrested in peaceful protest. As time has gone on many of those protesters facing the court system have been acquitted or the charges against them have been dropped.

Free speech and the right to demonstrate have been fought for and won in the UK. Independent journalists are a crucial part of our rights and the freedom to speak and share truthful information. But many large newspapers and media organisations in Britain and around the world are not independent. Some have made no effort or pretence of equality of ideas or opportunity. Subtle and not-so-subtle racism, homophobia and sexism are largely endemic in the British press. In terms of reporting environmental issues, they are mixed, with some papers being part of the solution and leading their readers along the path of a more truthful portrayal of events and others ignoring the situation and even misreporting the science.

The *Guardian*, the *Independent* and the *Daily Mirror* are good examples of the press that have all taken very powerful stances on the accurate reporting of environmental and climate emergency stories. The Guardian Newspaper Group took a pledge in 2019, telling us: 'We believe that the escalating climate crisis is the defining issue of our lifetimes and that the planet is in the grip of an emergency. We know that our readers and supporters around the world care passionately about this too, as so many of you have told us.' Their pledge includes a commitment to 'never be influenced by commercial or political interests and always be rooted in scientific fact'.

What can you do?

Take Action!

When you move to help someone you are invoking compassion. You want to do something kind or useful because you see suffering, loss or trauma.

When you speak up for animals who cannot speak, that is also a form of compassion. When you take action with compassion and love in your heart you will find the universe strains to help you, sends support and incredible solutions in all manner of ways. Miracles can happen: invoke compassion and service and see what unfolds.

Volunteer to help animals.

Help, support and speak up for people who cannot speak themselves.

Read and educate yourself about the environment and politics, new ideas and solutions.

Be proactive. Do something positive, meaningful, direct and useful.

SEE ALSO: Environmental Heroes and Inspirational People, p. 121; Environmental Law, p. 128; Group Action and Protest, p. 174

ADDITIVES IN FOOD

Do you know what you are eating? So many manufactured foods today contain additives, many of which have multiple effects on both the body and the environment. They are there to enhance flavours, to sweeten or to add colour.

Food additives can transform mushy raw ingredients into 'cheese-flavoured snacks' or 'tomato ketchup', to 'fruit flavoured' or 'chocolate flavouring'. Preservatives and antioxidants, mineral hydrocarbons and solvents all help make sure the food doesn't go rancid on the shelf.

Not all additives are bad for you, and not all are artificial. Some are natural and some nature identical, i.e. they're made to compare identically to a natural structure. Some are really necessary to preserve food. Additives are tested for safety and those given an E number are considered safe by the European Union.

Some additives, like the nitrates used to cure bacon and other red meats, have recently been found to be a major cause of cancer after years of industry argument. Nitrate-free meat products are now available in larger supermarkets and the market for additive-free meat is growing. Sulphur dioxide and sodium or potassium meta-bisulphite are some of the preservatives commonly used, known collectively as sulphites. Some people can get asthma and lung problems from sulphites.

Azodicarbonamide is a flour improver that is used to help bread dough stick together. People working in factories that make bread have higher rates of asthma and respiratory illness when they use this chemical. It was banned in plastic manufacture in 2005 by the EU but is still allowed in the USA.

Some food additives have been linked with asthma, eczema, rashes and watery eyes, as well as hyperactivity in children. Cochineal is a food colouring that is derived from the dried-up bodies of cochineal insects. Rare allergic reactions have taken place in some people eating cupcakes with these ingredients in the iced toppings.

Insects, however, are now used and promoted as a high

protein alternative to meat, and with many applications. Insects are raised without pesticides or antibiotics, and the industry says they can be raised humanely in cities without the need for land. New products are being created that use crickets, for example, to make protein-rich flour. Insects have been eaten as a staple protein across the world for centuries.

Food additives get into our bodies every day. When you eat heavily processed foods the chemicals reside in your body, and in the environment, through their production of toxic emissions and waste.

Here is a list of those additives that are most important to avoid for your overall health and the environment:

- **Sodium nitrite** – Fights bacteria in cooked meats and keeps them artificially pink. It has been found to change into molecules that cause cancer in certain conditions and increase heart disease.
- **Fructose corn syrup** – Increases the chance of obesity, diabetes, heart disease and cancer.
- **Food colourings: Yellow 5 & 6 and Red 40** – These are linked with cancer. Some food dyes are associated with hyperactivity, aggressive behaviour and learning difficulties in children.
- **Artificial sweeteners** – Studies show these can promote weight gain, obesity, cardiovascular disease and diabetes.
- **Carrageenan** – Some scientists say it causes digestive problems including irritable bowel disorder, inflammation and even colon cancer.
- **Guar gum** – Used as a laxative and for treating diarrhoea. It is also found in drinks and food as a thickener and binding agent.
- **Sodium benzoate** – Could cause obesity, allergies and inflammation.

Industry has invented some incredible solutions to the problems of keeping fresh foods on shelves longer. When we consider that 40 per cent of UK food is wasted the impetus for solutions is obvious. Food perishes according to the end of its season; for example, some fruits are only ripe and fresh for a few weeks in season such as strawberries, but consumers want to buy them all year long. Supermarkets want and need food to stay fresher longer not only on the shelf, but on picking, in transit and packing, and when you take it home. Scientists are now using more environmentally-friendly plant lipids, fatty acids and skin proteins to spray on food to increase the shelf life by up to three times its normal window for sale.

This new technology allows some products, like cucumbers, to be sold without a plastic wrapper as the spray makes a fatty seal around the food. Another example is tiny packets of chemicals which slow down the ability for the fruit and vegetables inside a box to rot. Fresh food gives off ethylene when it is breaking down, and attempts to reduce its effect on the rotting process have been successful. This technology stretches the season for several important fruits like avocados and strawberries.

Sugar is a powerful food additive and preservative. Sugar is found in many savoury dishes such as savoury pies, spare ribs, sausage rolls and table sauces and the vast majority of ready-made meals and comfort foods like chocolate, sweets and drinks. Of course, it's also found in milkshakes and soft drinks: an average 500 ml bottle of fizzy cola contains ten teaspoons of sugar. A standard size 330 ml can is likely to have 9.75 teaspoons of sugar, that's 39 grams per drink. NHS recommends a maximum of only 30g in a day for adults and 24g for children.

What can you do?

If there is to be compassion in the way we create our food, then crushing up large numbers of insects should be avoided. However small, each creature has the right to life. Ingredients that are humane and sustainable do not include insects or other animal body parts.

The easiest way to avoid additives and chemicals in your food is to eat real, organic or clean food. Avoid heavily processed items and cooked meats, salamis, processed cheeses, bacon and other red meats. Read the packaging, read the labels.

Decades of campaigning by individuals and groups have brought some success and clearer labelling laws so that consumers are better informed. So get smart, be aware of the ingredients and choose wisely.

Reduce your footprint on the earth by avoiding processed foods altogether. Eat fresh organic food.

Get more information from: The British Nutrition Foundation at www.nutrition.org.uk and allergy information from the Anaphylaxis Campaign at www.anaphylaxis.org.uk.

Eat a diet high in antioxidants, fresh vegetables and fruits. Build your resistance against illness by eating a healthy diet.

Avoid sugar when you can. Sugar is addictive, try living without it for three days or three months to see how deep you have gone. See how much power a grain of sugar has over you. Don't switch from sugar to honey, either. Let the bees eat their own food. Train your body to enjoy real food without over-sweetening.

More information is on this government website: www.food.gov.uk/safereating

SEE ALSO: Agriculture, p. 27; Allergies, p. 33; E Numbers, p. 119; Hormones, p. 196

AEROSOLS AND THE OZONE LAYER

The UK uses around 600 million aerosol cans each year, containing anything from deodorant and hairspray to paints and whipped cream.

In 1978 Sweden was the first country in the world to ban aerosol sprays that used chlorofluorocarbons (CFCs) as the propelling agent. Scientists had discovered that stratospheric ozone depletion was taking place. The ozone layer was depleting because the chemicals and by-products that included chlorine were reaching the delicate layer that protected us just eleven miles above our heads. The ozone layer protects everything on earth from the sun's dangerous ultraviolet rays. Nothing will survive without it. The ozone layer is a thin viscous membrane surrounding the earth that shields us. One atom of chlorine can destroy over 100,000 ozone molecules.

Campaigning by groups such as Friends of the Earth brought the issue to our attention and the British public campaigned, wrote letters and telephoned the relevant companies to tell them what they thought. The companies capitulated; they changed their manufacturing methods to remove the most dangerous chemicals.

In 1987 the Montreal Protocol was signed by 191 countries agreeing to phase out the production of CFCs and other damaging ozone chemicals.

While the most harmful of the chemicals were removed, somehow it didn't stop us from continuing to use the pressurised cans we know as aerosols.

Aerosols are literally the most expensive form of packaging available today. A complex layer of steel or tin, paper, plastic, all in a pressurised can. It's explosive and dangerous in many situations. It is only a package, however, a means of containing the liquid you require.

According to the British Aerosol Manufacturers' Association (BAMA), the UK now produces nearly 30 per cent of the aerosols in Europe and is third only to the USA and China in world production. Production has grown dramatically since 1990. About 1.4 billion aerosols are produced each year by companies in Britain and around 60–70 per cent are exported. UK expertise is recognised all over the world. Not all aerosols are dangerous to the ozone layer, but they are still highly explosive and dangerous to dispose of. The bestselling aerosols are deodorants and body sprays, but aerosols are not only used for products that you find around the home. There are around 2,000 brands of aerosol products on the market with over 200 different uses. Hospitals use aerosols to spray antibiotics on to wounds, and more recently to spray skin on to burn victims.

Asthma sufferers use aerosol asthma inhalers. Aerosols are used in industry, agriculture and science. Modern aerosols produce foams, mousses and creams, as well as the wet and dry sprays that we usually associate with an aerosol.

What can you do?

New designers and creative artists will be the eco saviours of the future. They will invent new forms of packaging and delivery of goods without this waste. Until then forgo aerosols and look for simple alternatives which are available for most items such as roll-on deodorants or liquid release tops, pressurised sprays and pump dispensers.

Be sure to recycle any aerosol cans you may use. Check if your local council collects aerosol cans in kerbside recycling. You can use the Recycle Now online tool to search for your nearest aerosol recycling plant.

Aerosol asthma inhalers can be recycled in some pharmacies via the GSK recycling scheme, set up by Complete the Cycle in

2011. Check for your nearest recycling centre on the Complete the Cycle online portal: www.pharmacyfinder.completethecycle.eu.

SEE ALSO: Allergies, p. 33; Deodorants and Antiperspirants, p. 95; Disposable Products, p. 105; Packaging, p. 240

AGRICULTURE

According to a report by Our World in Data, 50 per cent of the world's inhabitable land is used for agriculture. From this land, 77 per cent is used for livestock and 23 per cent for crops.

The methods of agriculture have developed over time in order to keep up with the growing population. Crops used to be grown only on land that had suitable soil and conditions. Small-scale farms used to produce a mixture of crops, and some of the land would be used to raise animals. It was worked carefully on a rota system, which would allow the land to rotate between different crops which used different minerals from the soil or fixed other nutrients from their roots.

However, because the need for food is so great, today's farming methods and especially the use of chemical fertilisers are having a destructive impact on the planet. Many farms now produce only one crop. This type of farming is called the 'monoculture system' and it comes with many disadvantages. The land becomes more easily exhausted because the same minerals from the soil are taken up by the crop year after year. Chemical substances are therefore used to provide the crop with the minerals it needs to survive. These chemicals are being used in large quantities to help unsuitable land produce crops.

These crops also become more vulnerable to pests and weeds and so exposure to weedkillers and pesticides is increasing due to constant overuse. Every year over 2.54 million tonnes of pesticides and weedkillers are used globally in industrial farming. These chemicals destroy ecosystems that used to help plants fight off pests; they are a major factor of biodiversity loss and extinction.

The spread of agribusiness – in the form of huge agricultural farms – is also harming the environment. Natural habitats are being destroyed; hedgerows are cut down to create huge open landscapes, for example. Losing the hedgerow is a catastrophic problem for local biodiversity and species are destroyed immediately. Our forests and rainforests are under threat too. Forests are being cleared in large areas. Current clearings are mainly for soya or palm oil plantations.

Soil is being overused and eroded. Deserts are spreading. Chemical fertilisers are polluting soil and seas, and as a result the number of different species of plants and animals in the world is decreasing.

While a quarter of our water supply is used to produce the food we eat, most food growing relies on rain falling. In the last few years especially, rainfall has become erratic, with droughts in some places leading to fires and too much rain in other places leading to flooding.

Through chemical farming, farmers are now producing surplus food in some areas which is either used as livestock feed or ends up in the landfill where it releases methane, which is a major greenhouse gas and poses a continuous threat to our planet. Poor distribution systems and unequal economic distribution mean that fresh food can be wasted in some places while other areas suffer from a lack.

What can you do?

Consider meat-free days. At least four times more land has to be given over to agriculture to feed people on meat than on vegetable produce. Consider buying organic food, which is healthier for both yourself and the land. Try to buy locally. Look out for your local farmers' market and buy seasonal foods. Grow your own food.

Become a vegetarian or vegan and live on a plant-based diet from your own grown vegetables or those from local farmers, markets.

Inform others. Contact WWOOF (World Wide Opportunities on Organic Farms) to learn more about organic farming. Contact Friends of the Earth, WRAP UK, the Soil Association and the Vegetarian Society, who have material on agriculture. Then try to get your school, college, workplace or library to mount an exhibition based on the information you have gathered.

SEE ALSO: Factory and Intensive Farming, p. 137; Litter, Waste and Landfill, p. 218; Veganism, Vegetarianism and the 'isms' of our Food, p. 308

AIR POLLUTION

The air in our major cities is polluted today with airborne chemicals and hazardous pollution from cars, industry and aeroplane travel. The World Health Organization says that nine out of ten people are breathing polluted air.

Air pollution causes infant deaths, respiratory and lung problems, heart diseases and long-term health problems like asthma. We have seen that worldwide, it kills seven million people every year, and causes early deaths in every age group. It is the fourth largest cause of death and ill health in humans after a bad diet, smoking and high blood pressure.

Indoor air pollution also poses a serious health risk. It comes from burning wood and coal for fires or cooking, mainly in developing countries, and from a mixture of chemicals found in furnishings, computers and innocent household items found in industrialised nations. Many people do not realise that deadly pollutants are also caused by frying food in the kitchen with cheap oil, or smoking in the home, from the paint on our walls, or from mould, damp, asbestos, formaldehyde and other contaminants such as non-flammable protection in our furniture, soft furnishings and curtains.

According to the European Environment Agency, transport emissions in the EU are responsible for 10 per cent of total air pollution and more than two thirds of all NO_x (nitrogen oxides) emissions although the future trend is downwards. Battery electric cars and hybrids are increasing in choice and design, and sales will overtake those vehicles with the internal combustion engine in the years to come.

The combustion of fossil fuels – oil, coal and gas – in industry contributes directly to air pollution and the climate emergency. Other contributors include poisonous gases, smoke and ash from wildfires, dust and household pollution, farming with chemical sprays, smoke and chemicals in the air from other industries such as concrete production and its use on building sites.

Regulation is the key to achieving cleaner air, using laws and government policies to tell industries to create cleaner energy, to reduce their emissions and to ban highly toxic cars and other vehicles. The Kyoto Protocol was adopted in Japan in 1997, and signed by 183 countries which agreed to reduce their carbon dioxide emissions. The United States of America is the only industrial giant which has refused to sign it and has withdrawn from the Paris Agreement on reductions of carbon dioxide too. In 2006 the World Health Organization created new air quality guidelines to reduce air-pollution-related deaths in humans by 15 per cent year on year.

Some tiny pollutants known as particulates, like PM10

and PM2.5 (which relates to their size, e.g. 10 micrometres of pollution in diameter), are unseen to the human eye and extremely dangerous. They are small enough to pass through the body and damage our organs. People can die from ingesting these pollutants in tiny amounts. There is growing concern that vehicle and industrial pollution is so dangerous that there should be a worldwide ban on certain chemicals.

Some pollutants even affect children's brains and their ability to learn and study and their educational achievements. Poor air quality has also been proved to lower a child's IQ. These toxins are more harmful to children than adults as their brains are developing.

The British Government says that the impact of air pollution on humans, their health and welfare, and on the natural environment is so serious it must be measured and reported. The quality of the air in Britain is currently so bad that each year it is estimated that 110,000 people or more die prematurely from air-pollution-related diseases. Air pollution directly aggravates cancer, diabetes and dangerous lung diseases in vulnerable people.

In 2020 the World Health Organization said that high incidence of air pollution could raise the risk of dying from Covid-19. Early studies have suggested that areas with high incidence of polluting air particles have also seen a similarly high number of patients with Covid-19 and patients dying with the disease.

Air pollution is directly linked to the climate emergency. The World Health Organization says we need to use cleaner burning technologies and cooking fuel, avoid burning waste and recycle as much as possible.

Many governments are trying to solve the problem of air pollution by burning fossil fuels more cleanly, by making it mandatory to fit catalytic converters to new cars, by encouraging people to drive electric cars fuelled by renewable energy sources, and by using alternative, less polluting fuels.

In 2020 during the coronavirus pandemic lockdowns, it was estimated by the World Economic Forum that more than 11,000 lives were saved in Europe by the reduction in air pollution in this very short period. The plummeting levels of pollution resulted from a 95 per cent reduction in car and aeroplane use. Figures from the Children's Hospital of Philadelphia (CHOP) in the US also showed that 6,000 fewer children developed life-threatening asthma emergencies than would be expected in the same thirty-day time period and doctors saw fewer cases of chronic obstructive pulmonary disease (COPD). We had a rare glimpse of what a world could look like when our choice of transport fuel did not kill us and restrict our breathing.

If our transport systems were pollution-free and carbon neutral we could all have our private cars, and even fly across the planet. Our lung and heart health, and our general health and community systems, and essential infrastructure and the environment would benefit enormously. Less pollution equals less death and distress from polluted air.

What can you do?

The air around you is as precious as the water you drink. Your breath is the one thing that means you are truly alive.

Make sure your home environment is clean and clear, with fresh air when possible. Use high quality masks and filters for airborne chemicals and damaged, smoky, or otherwise dirty air. Cut down on burning oils and fried foods at home, clean up moulds and open your windows every single day.

Don't add to the pollution and low air quality yourself.

Instead, cycle, walk, take public transport and use electric vehicles.

Don't burn leaves and clippings in your garden and add to the problem, compost garden waste instead.

Call governments to account. They must value people's lives and health over speed and the continued burning of fossil fuels for industrial processes.

There are many organisations working towards cleaner air that have created valuable reading materials, such as the World Health Organization, and the UK Government Air Quality Report.

Organisations working on Air Pollution issues include: Friends of the Earth, the Healthy Air Campaign and the British Lung Foundation Campaign for Cleaner Air. Join them, give them your support, write letters and be active.

SEE ALSO: Climate Emergency, p. 64; Fossil Fuels, p. 160; One Planet Living, p. 234; Transport, p. 294; Your Carbon Footprint, p. 322

ALLERGIES

One in five people is likely to suffer from some sort of allergy. If you have ever suffered from hay fever, you will know how uncomfortable that can be. Allergies can seriously affect a sufferer's way of life, restricting everyday activities such as eating out or visiting friends with pets.

An allergy occurs when the human body becomes sensitive to a particular substance or chemical. The body's immune system thinks it is being attacked and produces more and more antibodies, which include powerful natural chemicals like histamine. Histamines help the body get rid of foreign substances that irritate it so might make you sneeze for example. These histamines provoke physical reactions: asthma, eczema, watering eyes or a runny nose. Allergic reactions can be so extreme that they result in death, though this is rare. Antihistamines are used to reduce the effect of the allergy symptoms on the body.

Some of the chemicals in our environment could be triggering allergic reactions in our bodies. Pesticides, beauty aids, detergents and food additives are only some of the chemicals contributing to our bodies' daily dose of foreign substances.

The chemicals can also produce a 'hidden allergy', where the body does not immediately react against a substance in visible ways. Instead, hay fever responses give way to long-term illnesses like depression and migraines. What's more, with chemicals all around us, many people are becoming sensitive to the overload of too many allergens at once. This, too, can trigger an allergic reaction.

Allergies are difficult to treat. Although medicine can relieve the symptoms in the short term, it cannot treat the cause. Only by eliminating chemicals from our bodies as much as possible will we be able to cure allergies in the long run.

What can you do?

Allergic reactions take place when the body is stressed. Mental and emotional stress compound the body's sensitive reaction to outside stimulus. The core of you is stressed if you are reacting to a number of things in an allergic way. Tackle that as a priority.

How can you de-stress and reduce your reaction to outside pressures? How can you make calming, healing, loving yourself and nurturing yourself be your main job while you get strong? Look at the pressure areas in your life. If you can, try to reduce commitments and concentrate on your health for a while. Reduce or avoid caffeine, chocolate and alcohol which create more stress and tiredness in the body. Increase your intake of fresh green vegetables, maybe drink green smoothies.

Increase your time spent outdoors in nature as much as possible.

Become more informed about the artificial chemicals you are putting into your body, and see what you can avoid. The fact that a given product is marked as safe only means that in trials there were no unacceptable short-term side effects – it says nothing about the long-term effects or the cumulative effects of many of these chemicals being used together.

If you think you may suffer from allergies, visit a doctor or homeopath, or try self-help by keeping a record of what you have been eating or been close to. You may discover the cause of your allergy yourself.

Allergy UK and Action Against Allergy provide information, advice and support on allergies.

SEE ALSO: Additives in Food, p. 20; Chemicals, p. 54; Hormones, p. 196

ANIMALS AND INSECTS IN THE COSMETICS INDUSTRY

Animal testing in the cosmetic industry has long been a troubling issue and one that many campaigners have rightly brought to the fore. What is less well-known is the industry's reliance on animals and insects as ingredients of its products.

From foundations and eyeshadows to moisturisers and lip gloss, the core component or the active ingredient of a product that we take for granted is often derived from some part of an animal. It is very possible that your moisturising cream includes pig's urine or that your red eyeshadow and lipstick is made of crushed insects.

Some modern anti-ageing creams for skin contain the mucus and slimy secretions of snails. They are often marketed and sold as an anti-acne treatment. Mink oil from purified carcass fat is added to cosmetics to make them soft on the skin.

Some of the following living beings are being exploited, and organisations such as animal rights group PETA are fighting for the end to cruel practice.

Animal hair

The brushes we use on our teeth and hair, or to apply make-up or shaving cream, unless synthetic, can be made from animal hair sourced from ox, sable, monkeys, ponies, horses, goats, mink, hogs and squirrels. It is known that some manufacturers are using real animal hair for their brushes and are sourcing the materials from countries where animal rights are either poorly regulated or even unregulated.

Advanced technology has made it possible to create high-quality synthetic brushes that hold colour and resist shedding, although plastic alternative bristles include agave cactus (Tampico) and the Indian palm tree (Palmyra).

Beavers and Civets

A fixative which is commonly found in our fragrances and perfumes is the by-product of otter, beaver, musk deer, and civet cat genitals. The gland very near the genital organs is scraped to extract the valuable oil. The animal rights charity PETA have claimed that caged civet cats are being whipped in the genitals to activate their sexual scent. They say the musk-oil industry also involves shooting deer and trapping beavers.

Factory-farmed animals' skin, bones and tendons

Gelatin is a protein gel obtained by boiling the skin, tendons, ligaments or bones of cows, pigs, horses and fish. It is used in nail strengthener, nail-polish remover, shampoo, face masks and cream cosmetics. Gelatin can also be found in food products such as desserts and sweets.

Fish

Another common cosmetics ingredient is guanine, a light-diffusing crystalline component. It is found in crushed fish shells, which are used to create a shimmery effect in eyeshadows, eyeliners, nail and lip products and skin-lightening creams. It is also used in the cleaning industry and can be found in many common bath products.

Horses

Oestrogen for use in conventional cosmetic products can be derived from female hormones found in pregnant horses' urine. It is commonly used in creams and perfumes, in pharmaceutical drugs for reproductive problems and in birth-control pills and menopausal hormone pills.

Insects and Bees

Carmine, Crimson Lake, or E120, is a red pigment used not only in beauty products, but also food colouring, paint and textile dyes and it comes from the female cochineal beetle which is boiled, filtered and mixed with different substances to achieve varying shades of red. Carmine can be found in powdered beauty products such as blushers, eyeshadows, lipsticks and nail polish. According to animal rights group PETA, it takes 70,000 crushed female beetles to get 454 grams of red dye.

Bees are one of the most important species on earth. Their honey and beeswax is commonly used in the cosmetics industry in lotions, sunscreen, skin creams and bath crèmes as an emollient, a softener. An average worker bee makes about one twelfth of a teaspoon of honey in its entire lifetime.

Sharks

Squalene is a type of oil that mimics the consistency of the human sebum, an oily substance that is produced by the sebaceous glands – microscopic glands – which naturally keep the skin and hair moisturised. While it is often derived from plants, the highest concentration of squalene is found in the oil of shark liver. The squalene keeps the creatures buoyant in the water. Ninety per cent of shark-liver oil industrial use is for lip balm, lipsticks, sunscreen and cosmetic products which require its water-soluble and moisturising properties according to a 2015 report published by Bloom Association, a French ocean-conservation organisation based in Paris. It also has anti-inflammatory ingredients and is used in several medicinal and pharmaceutical products.

Many deep-sea sharks are already endangered. Bloom found that at least six million sharks are killed each year for their squalene. They tested seventy-two products, all of which contained squalene.

Sheep and Cows

Lanolin, an ingredient widely used for its moisturising properties especially for dry skin, is a thick emulsifying wax produced by sheep. It is extracted from the sheared wool taken for the textiles industry. Lanolin is commonly found in imported soaps, creams, mascara, lip products and nail strengtheners.

Tallow, from sheep or cows, is a main ingredient in soap bars and cleansers too, made by boiling their fat and bones with sodium hydroxide. You will find tallow labelled as hydrolysed collagen in many anti-ageing skin creams as well as soap bars, lip products, foundation, primers, deodorants, shaving creams, toothpaste and hairsprays.

Allantoin can be found in skin-conditioning agents such as creams, deodorants, sunscreen, lipstick and hair-dye. It is often plant derived, but when animal derived, it is obtained from the uric acid found in cow urine.

Some of this is well known and there is nothing wrong with it either, so long as the sheep and cows are reared properly.

Compassion in World Farming point to lameness, mutilations and illness being a pattern in sheep rearing of some farmers with poor standards. They remind us that several million sheep, pigs and cows are kept permanently indoors and will never have access to pasture to graze.

Sperm Whales

Sperm whales are carriers of the most sought-after fixative in the perfume industry – ambergris – which is a waxy substance excreted from their intestines. It can be found floating on the ocean surface where it is maturing its scent, or on beaches as a dark stone-like substance. It is illegal to use in some countries including the United States but is still used elsewhere.

What can you do?

Avoid animal products in your cosmetics; make them yourself, or look out for certifications such as: Certified Vegan by Vegan

Action, Vegan by The Vegan Society, and Vegan Approved by The Vegetarian Society and Cruelty Free International. These groups list cosmetics that use plant and vegetable alternatives.

Take stock of all your cosmetics and products around the house. Make a list of their recurring ingredients and check them for hidden animal ingredients.

The link between agriculture and the overuse of animal by-products in other industries means we need to stop the chain between them. Awareness of product ingredients is a vital step.

The cosmetics company Lush has been running a campaign against using shark oil and sells Shark Fin Soap with profits going to the Rob Stewart Sharkwater Foundation, a charity set up after Rob Stewart died while filming a documentary about killing sharks for their fins.

SEE ALSO: Agriculture, p. 27; Animals in the Textile Industry, below; Animal Testing, p. 45; Cosmetics and Toiletries, p. 80; Hair Products and Hair Accessories, p. 183; Hormones, p. 196; Tropical Rainforests, p. 304

ANIMALS IN THE TEXTILE INDUSTRY

Some people will kill for fashion, literally. Every year, billions of animals suffer horrendous torture and death to service the fashion industry.

Animals die to provide skin for leather goods. According to undercover reports and campaigns by animal rights charity PETA, birds are held under force whilst their feathers are ripped out of their bodies to be used for feather linings. Their special investigation shows horrible cruelty on goose farms with live plucking used in China and Hungary where the majority of our feather down comes from. Small animals are caged before being killed for their fur. Sheep have been found beaten and mutilated by workers who shear their wool in the UK. Goats have been mistreated for their cashmere and mohair.

The vast majority of the world's textile production is in India, Bangladesh and China. Not all countries have laws regarding animal welfare and manufacturing processes, including the disposal of toxic chemicals. Factory workers and people living near textile factories in polluted areas are at risk of health problems such as skin diseases and respiratory illnesses, and people die from exposure to high concentrations of dangerous chemicals.

Down

Down – feathers from birds such as ducks, geese and swans – is favoured in fashion, outdoor wear and homewares for its warming properties, especially against below-freezing temperatures. Due to high demands for down, farmers usually pluck from live birds in order for them to produce more feathers in the future. It is a very painful experience for the bird.

Some suppliers hold the Responsible Down Standard (RDS) certification; however, as an article published in the *Economist* pointed out, 80 per cent of the world's down is produced in China, where there are no animal welfare laws.

Fur and Exotic Animal Skins

The ethics surrounding the sourcing and selling of fur for use in the fashion industry are widely debated. In 2003 the UK joined other countries in banning fur farming; however, it is still legal to sell fur products in Britain. The industry markets fur, hair and skin as 'luxury' including that from rabbits, minks, lynxes, racoons, coyotes, wolves, beavers, possums, chinchillas, goats, foxes, alpacas and llamas. Exotic animals used include reptiles, snakes, sting rays, emus, kangaroos and horses. Not only are these animals exploited for their luxury properties but, additionally, some of the 'exotic' animals are at risk of extinction.

The fur and skins industry also subjects animals to inhumane treatment. In 2014 a Chinese angora factory was exposed with video footage showing angora rabbits chained up and being skinned alive. The footage caused widespread outrage and led some retailers and designer houses to discontinue the sale of angora products.

Faux fur and skins are a good alternative; however, at times they are produced to such a high standard that it is difficult to tell whether they are real or not. This has led to incorrect labelling as faux fur when in fact it is real fur. Furthermore, faux alternatives are usually made from petroleum-based polyester and nylon which, if irresponsibly disposed of, can end up in landfill where they can take decades to degrade, and furthermore will degrade into microplastics.

Leather

Cattle, the greatest source of leather, require a huge amount of arable land, and often wholesale deforestation has taken place to create grazing for these animals.

According to Science Direct, the dyeing process of leather – tanning – requires around 300kg of chemicals for every 900kg of skin.

Be aware that some alternatives to leather, known as vegan leather or pleather, are commonly made from polyvinyl chloride (PVC) plastic, which Greenpeace lists as one of the environment's most damaging plastics. However, there are ethical, sustainable, vegan alternatives available and these are becoming more and more popular, including mushroom leather.

Silk

Silk has been known as a timeless, elegant, luxury fashion fabric for thousands of years and is favoured in the fashion industry for its flame-retardant, climate adaptation and anti-bacterial properties. It is made up of threads that form inside the cocoon of the mulberry silkworm, a process which produces around 90 per cent of commercial silk. The threads are formed from the silkworm's saliva. The process of extracting the threads is controversial because it involves boiling the cocoon while the insect, or pupa, is inside, in order that the strands remain intact.

Silk harvesting is a low-waste process. The silkworm's diet is simply the leaves from the mulberry tree – a robust tree that has many other uses. What's left of the silkworm and cocoon doesn't go to waste either: the pupae are a rich source of protein, making them a popular snack across many Asian countries, and the outer part of the cocoon is used as a natural fertiliser.

Look for natural silks such as raw, undyed and unbleached, or 100 per cent naturally dyed silk. Choose OTEX-certified organic silk, or ahimsa silk by Cocoon, which is GOTS-certified (Global Organic Textile Standard).

Wool

Wool, because it is a natural fibre, is breathable and biodegradable.

In theory sheep shearing for wool does not have to be cruel. It might even be considered good animal husbandry to shear a sheep during the summer months in order for it to be kept cool, but the wool industry is under scrutiny over animal welfare.

There are growing concerns regarding the conditions

in which animals are kept and handled during shearing and mulesing – a process used to control flystrike which, if untreated, can be fatal. However, mulesing which involves 'marking' the lamb by cutting crescent-shaped flaps of skin from around a lamb's breech is a process which is criticised by the RSPCA. They believe it is unacceptable to breed sheep which will require this cruel treatment and urge retailers to buy wool from non-mulesed sheep.

The demand for wool is high and one of the results is intensive sheep farming. This method harms the environment: when sheep overgraze an area, the land rapidly degrades. Millions of sheep are also kept for their whole lives indoors in huge agricultural sheds and warehouses.

Popular fashion houses and brands are starting to listen to consumers; they are becoming aware of the impact that stocking clothing made from exotic animals and fur is having on their reputation. This is a step in the right direction for the welfare of these beautiful animals.

However, there is debate around the use of animal products. Fur Europe works to inform about the sustainability characteristics of fur.

Programmes such as WelFur and Fur Europe's Open Farm scheme, both of which grant first-hand insights into the industry's animal welfare standards, are needed to provide clarity on what is an understandably emotional issue. Only then can consumers make an informed decision.

One thing to remember is that supply chains are complicated. While a company may carry out the necessary checks on the farms from which they are sourcing their fibres or the factories where their products are being manufactured, it's essential that they remain vigilant and regularly check up on their long chain of suppliers.

What can you do?

If you decide to purchase an item derived from an animal, first look for second-hand items or recycled fibres. If buying new, the accreditations to look out for are: Responsible Wool Standard, ZQ Merino Standard, the Soil Association Organic Standards, European Commission's Welfare Quality and the Responsible Down Standard (RDS).

Follow organisations such as the Good On You app, Fashion for Good, Fashion Revolution, and the Clean Clothes Campaign. See the campaigns by Beauty Without Cruelty and PETA. These organisations and campaigns work to raise awareness of brands' sustainability and campaign for environmental and social practices within the fashion and textile industry.

SEE ALSO: Agriculture, p. 27; Fashion and Clothing, p. 139; Species Extinction, p. 285; Sustainable Textiles, p. 288

ANIMAL TESTING

After forty years of campaigning and education about the cruelty of animal testing, there are still companies in the UK and around the world that test cosmetics, soaps, perfumes, household products and many other everyday products on animals.

There is a ban on such testing of cosmetics and make-up produced or sold within the EU, but this is not the case in the USA and China, for example. In fact, around 115 million animals around the world are used every year in tests for research, products, medicines and chemicals, as reported by Humane Society International.

Rabbits, mice, dogs, monkeys, guinea pigs, cats and other animals are all given household products like soap to see whether they produce an allergic reaction, whether they make their eyes water or sting, or whether they make them sneeze. Animals are used before a product is marketed to us as 'safe'. This is cruel and barbaric. Also the tests are mostly unnecessary as human testers would give more accurate results and more specific and detailed information. A horse or dog cannot say that a product stings or that a rash is sore – only human testers can really do that. Some companies have always used human testers, for example Neal's Yard Remedies, and some, like the Body Shop, were set up with the pledge that they would never use animal testing.

Nearly four million tests were carried out in 2018 on living animals in the UK alone according to Home Office statistics. About half were used for experiments, testing and scientific research for products. The other half – about two million animals – were used in breeding and DNA experiments, including the breeding of genetically engineered animals.

Most of the experiments are carried out on mice, fish and rats. As a result of campaigning and a public outcry, horses, monkeys, cats and dogs are now classed as 'specially protected species' and any scientific laboratories must give really valid biological reasons to include them in their research. The BBC reported that in the UK during 2017 some 2,215 monkeys, 2,496 dogs, 228 horses and 71 cats were experimented on for research that met those criteria.

UK law states than when a company has found an alternative to animal testing for a drug or product, they must use that.

If your product comes from abroad it is much harder to enforce laws and check the status. Buying cruelty-free cosmetics and household cleaners is the only way to be sure that there is no animal testing or animal ingredients in those products.

The killing of animals for human use is not new. Our ancestors often killed animals to honour their gods. Killing without conscious awareness of their suffering affects all involved. This is why our ancestors carried out rituals for animals killed in hunting.

An overview of our relationship with the animal kingdom leaves us humans wanting. We may consider ourselves an evolved species but other sentient beings are being murdered for our progress. There is suffering, pain and loss.

What can you do?

Do we need to be part of this suffering? Is your unconsciousness of this suffering allowing it to continue? Do animals have a right not to be part of this research? What has developed in us as humans that we continue to allow this suffering?

The medical students and doctors of tomorrow already want to change. They see that research by humans for humans is possible. Science will create a solution, especially if business really needs it and has no exemptions for animal testing. The National Anti-Vivisection Society offers help to students who do not wish to use animals in biology lessons and research.

Hold a funeral service or a ritual of remembrance, write a prayer or a poem to remember the animals that have been killed for your development. Honour them in some way.

Buy products that are not tested on animals – double-check on their information and marketing material. Look out for simple, pure alternatives to treat minor problems and aches and pains, like plant-based medicines and homeopathic remedies (not instead of a doctor – obviously, seek medical advice if you need it).

Contact animal rights groups for more information, advice and product lists.

Join and support animal advocacy groups and charities that

are working to protect animals: Cruelty Free International Trust, Animal Free Research UK, The Coalition for Consumer Information on Cosmetics – Leaping Bunny Programme, Viva!, PETA.

SEE ALSO: Animals and Insects in the Cosmetics Industry, p. 35; Cosmetics and Toiletries, p. 80; Species Extinction, p. 285

B

BATTERIES, RECHARGERS AND INSTANT ENERGY

Batteries are dangerous and powerful little packets of energy. We use millions of them to power our phones, computers, gadgets and all manner of technological inventions, but do we understand the impact they have on our environment? The chemicals used in batteries – mercury, lithium, zinc, lead and cadmium – are actually extremely toxic and if they are left in the rubbish, in landfill sites, these chemicals leach and are absorbed in the soil.

In Tibet, lithium mining has caused tremendous environmental destruction. Protesters have shown dead fish from the toxic chemical leaks from the Ganzizhou Rongda Lithium mine, which has been destroying the surrounding ecosystem. Local people who live on very little and cause almost no environmental impact have seen dead cows, dead yaks, dead fish and other animals – all contaminated from the mining activity directly connected to creating and supplying batteries for smartphones and electric cars.

Bolivia, Chile and Argentina produce about half the

world's supply of lithium batteries. These hot dry countries have found that the lithium mining businesses are using 500,000 gallons of water for every tonne of lithium they extract. This has a massive impact on local farmers, most of whom grow food.

When batteries are produced the impacts on fish and wildlife downstream of the factories and processing plants can be seen up to 150 miles away. Deep-sea mining for lead also causes a loss of marine species through the release of toxic chemicals.

Amnesty International has called out battery manufacturers and electric car manufacturers for procuring batteries from unregulated industries whose production methods have affected human rights and caused environmental impacts. They have documented terrible conditions in the Democratic Republic of the Congo, for example, including child labour and workers being exposed to health risks from the contaminants and toxic chemicals used in the production of batteries. They also point to the lack of information given to local people in countries like Argentina, who often agree in the first instance to have a factory only to discover, too late, that their water sources have been contaminated.

What can you do?

Batteries are a sign of a quick-fix life. What you need is to hear the radio or music without batteries being involved in the first place.

The inventors and designers of the future are born already, thinking and working on designs that will make batteries obsolete. They will create many new ways to make energy for games, phones and things we love to work with.

We will be collecting batteries and recycling their compacted energy for generations to come, even when their technology is defunct and new forms have taken over.

Reduce your need for products with batteries. Treat them as something for exceptional circumstances not the rule. Use rechargeable batteries as a last resort.

Never throw batteries in the landfill and general waste stream. They must be taken separately to a battery waste processing centre, or to a community waste-collection point. All shops that sell large amounts of batteries must provide recycling collection points under EU law. Councils will have an online list.

Amnesty International wants companies who use batteries to produce data and information about the environmental impact of their supply chain. They have urged the electric vehicle industry especially to come up with an ethical and clean battery. Electric vehicles are set to increase in numbers exponentially in the next ten years.

Look for current alternatives: wind-up machines like wind-up radios; solar-powered battery and telephone chargers; bicycle-powered energy and any alternative that doesn't need batteries at all. Bicycle-powered appliances are increasingly available, including washing machines, food processors, music centres and public address systems; there are even bicycle generators that could power your whole home.

You can't have it all. Why should children in Africa and South America have no clean water to drink because you have so many products that use dangerous chemicals for their power source?

SEE ALSO: Consumption and Consumerism, p. 77; Disposable Products, p. 105; Energy Efficiency and Energy Conservation, p. 116

BIODEGRADABILITY AND THE ENVIRONMENT

The term 'biodegradable' is used to describe the way that substances break down in the environment. For example, a paper bag will break down over time into a collection of harmless substances like water, air and carbon. These substances can then be reabsorbed into the natural soil, wind and water cycles of the planet.

It is a major problem that many of the products we use are 'non-biodegradable'. This means that they take an enormously long time – sometimes thousands of years – to break down in the environment and they can cause damage while they are still whole. What's more, the substances produced when, for example, fertilisers break down can be even more dangerous than the original fertilisers themselves.

Plastic can last in the soil for thousands of years. Some shops and manufacturers are now labelling certain products, like plastic bags, as biodegradable, but unfortunately most are only 'biodestructible'. This means that the plastic breaks down in the environment to tiny particles known as 'microplastics' that the human eye cannot see.

So the problem isn't necessarily solved – it's just put off. In fact, microplastics bring issues of their own – they are dangerous to small animals and insects that can ingest the tiny particles. Truly biodegradable plastic is extremely expensive and even the British Plastics Federation doubt it can be used in every circumstance.

What can you do?

In the long term everything eventually disintegrates. Mother Earth absorbs everything; even stone and rock are part of her and the inevitable disintegration and reabsorption takes place no matter what. The natural world's carbon cycle occurs within a long timescale, yet within a short period human beings' toxic actions have upset the balance and permanence of the soil and air around us.

If you live lightly on the earth you try to leave it clean and as you found it at the very least.

Cut down the amount of disposable products you use. Compost your peelings and eggshells and grow your own food. Keep the cycle of food and waste within your own small ecosystem and cut out waste from packaging altogether.

Read the section on 'Plastics' for examples of materials that take a long time to degrade and can be dangerous.

Try to buy products that will biodegrade easily; choose paper, for example, which can be recycled, rather than plastic products. Avoid aerosols, which do not biodegrade.

For more information, go to organisations such as WRAP UK, This is Rubbish, Food Cycle, FareShare, FoodSave, Fruit Factory and Waste Aid.

SEE ALSO: Disposable Products, p. 105; Litter, Waste and Landfill, p. 218; Plastics, p. 253

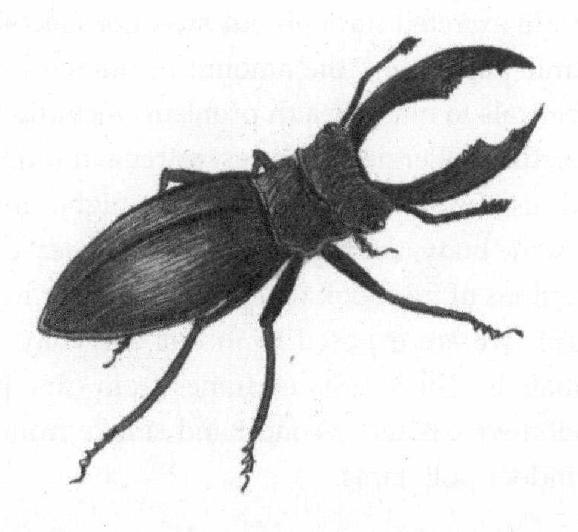

C

CHEMICALS

Our bodies contain hundreds of different chemicals all working together, keeping us alive. However, many aspects of our everyday living – what we eat, who we surround ourselves with, how we travel – contribute to the artificial chemicals we are adding to our bodies in tiny but accumulating amounts.

Our bodies do not always know what to do with these chemicals. Some are stored, causing organ problems later; some clog our bodies with toxins, such as in the lymph or thyroid glands, and others are excreted through our sweat or faeces. Medical and academic papers link the amount of human toxic exposure to chemicals to many health problems including eczema, autism, infertility, allergies, diabetes, asthma, multiple sclerosis, thyroid disease and heart disease. The higher the level of chemicals in the body, the higher the level of disease found.

Other sections of this book will provide further insight into the chemicals we are exposed to in our everyday activities, such as inhaling vehicle exhaust fumes from cars, pesticides from the food we eat and second-hand smoke from smokers and other indoor pollutants.

Every year around one million artificial chemicals are manufactured; thoroughly testing them all is proving difficult. During chemical testing, industry is not obliged to produce full enough results for proper analysis, and some chemicals may not even be tested at all. Furthermore, tests are not done to see the cumulative effect on the human blood and organs of different household and personal body items used in combination; they are tested as if this is the only product you will use. Each company is only responsible for their product, although it may be part of a potentially huge soup of chemicals that you are exposed to every day.

The human body has a sophisticated system of sorting out and storing good substances and in tiny amounts. Yet however brilliant, it can't always differentiate artificial chemicals from natural ones found in our blood and so they are also stored in our bodies. This leads to synthetic chemicals building up in our bodies and blood, mostly in our fatty tissue. They overload the system and it breaks down, allowing mutations and viruses to become embedded in the fatty tissue. This is one reason why so many women today are developing breast cancers.

The fatty tissue surrounding the breast is the first place the tiny particles of chemicals from underarm sprays and antiperspirants might seek. Trying to solve one problem, sweating, often leads to the creation of another as ducts and pores that should be excreting toxins naturally through sweat in the armpits become blocked.

The modern world has exponentially increased the risk of cancers of all kinds. In 1962, Rachel Carson, a marine biologist, wrote an earth-shattering book called *Silent Spring*. It was a chillingly realistic account of how small amounts of chemical poisons like pesticides get into the rivers and streams and water supply and bio-accumulate, building up over time to a toxic level that will fatally damage the ecosystem.

She denounced the widespread use of chemicals in household

products and in industry and the lack of care in how they were disposed of through the main water systems and into the oceans: 'Even in the vast and mysterious reaches of the sea we are brought back to the fundamental truth that nothing lives to itself.' Rachel Carson knew that all of life was interconnected, that the waters of the earth are also part of us and polluting them would have a long-term effect on our health and the health of every other living creature.

Dioxins, furans and organochlorine chemicals are one of the most dangerous and highly toxic clusters of chemicals known. They are produced as unwanted by-products of the use of chlorine in industrial smelting, chlorine bleaching in wood pulping and paper manufacture, in the manufacture of pesticides and herbicides, from uncontrolled and incomplete burning processes in incineration and from PCB-based industrial waste oils.

They have bio-accumulated in our bodies and our food chain. Most of our exposure is through animal fats in meat and dairy products and in fish and shellfish especially. They are dangerous, not just because of their chemical toxicity but because they are stable chemicals, able to be absorbed and stored by human fatty tissue for many years. The toxicity of dioxins is thought to be sixty thousand times higher than potassium cyanide.

Dioxin traces have been found in some mothers' fatty breast tissue and breast milk along with DDT, a powerful, artificially made chemical pesticide. These stored chemicals can cause harm later in life. Women do not deliberately spray themselves with DDT; they are exposed over time through their lifestyles and diets, through air pollution and through the water. Most people have some level of dioxin contamination in their fatty tissue. The developing and growing foetus in the womb is the most sensitive to dioxin exposure and a newborn baby with fast-growing hormone systems is thought to be more vulnerable to particular effects.

Chemicals in our food build up and accumulate in our blood and organs. We become what we eat. When the human body is flooded with unnatural particles, sometimes blood cells and good antibodies fight or send misleading messages to the organs and immune system, causing problematic illnesses. Sometimes our immune system is damaged or weakened. The lymphatic system is our first line of defence against disease.

Lymph is a fluid that circulates through the body in the lymph vessels. It drains excess fluids and proteins from the tissues all around the body, and sends them back into the bloodstream. The lymph nodes also remove waste produced by blood cells and have a role in fighting infections, bacteria and viruses. In the lymph nodes there are small glands that produce defensive chemicals to attack and kill dangerous threats to the body. A build-up of artificial chemicals in the blood can disrupt their messaging, sometimes blocking their ability to work and causing cancers to occur.

There is substantial evidence to connect our high artificial chemical intake with the high incidences of cancers in society. There is no escaping these chemicals completely in our modern life but knowledge and information could keep us safer and lessen our own exposure. Industrial chemical production has been doubling about every ten years and the problem is growing.

What can you do?

Chemicals are part of us. We have manufactured and created them and we have to be responsible for them. Chemical overload affects humans and other species too. Your household products affect the ecosystem – both where they were made and where we dispose of them.

Don't spill your waste in the beauty of your river. Your

route to the sea starts in your sink and toilet. Your route to the sea starts with the gullies and gutters in the streets outside your home.

Chemicals kill and destroy small ecosystems, stream by stream, river by river. The number of rivers polluted by spills, emergencies, outflow from business and industry grows.

Avoid using artificial chemicals whenever you can. Exercise regularly to allow toxins and unwanted chemicals to be released naturally through sweating in the body.

Try buying organic food and clothes, avoiding clothes that need dry-cleaning. Find environmentally-friendly alternatives for cosmetics and toiletries and cleaning products that are high in artificial chemicals. Breast Cancer UK has lists of household products, cosmetics and other beauty items that are chemical-free. Read the labels on what you buy. If a warning label says that a substance is unsafe for pets and children, it is logical to assume that it won't do you much good either.

Be careful how you dispose of chemicals, aerosols, batteries, medicines, household cleaners and especially bleaches and other dangerous household items.

If you want to find out more about specific chemicals, contact the London Hazards Centre. Trade unions and organisations such as the London Hazards Centre also help people affected by chemical hazards at work and in the home.

SEE ALSO: Acid Rain, p. 15; Air Pollution, p. 29; Chlorine, p. 58; Fossil Fuels, p. 160; Hormones, p. 196; Pesticides, Herbicides and Agri-Chemicals, p. 244

CHLORINE

Chlorine is found naturally in salt (sodium chloride), either dissolved in the sea or in rock form and as a natural gas.

In its elemental gas form, chlorine is always human-made. It is fundamentally the most important chemical that is used to kill bacteria and viruses in water. Chlorine in local water systems has been proved to reduce diarrhoea and in many countries drinking water is lightly chlorinated. Chlorine is also used in swimming pools as a disinfectant, allowing us to swim safely. It is used in people's homes as a disinfectant and as a fabric whitener. It is an important antiseptic.

Household bleach is chlorine in the form of sodium hypochlorite. This will give off chlorine gas if mixed with other household cleaners or detergents and is potentially harmful.

In its gas form chlorine is toxic and a dangerous irritant. It has been used as a weapon in wars since 1915, to indiscriminately kill and harm people, crops and animals. In large amounts it attacks the lungs, throat and eyes, which are damaged if the victim does avoid death. One must hope that leaders will never again use chlorine gas as a chemical weapon on their own citizens or others. Unfortunately, the last time chlorine gas was used in a war was as recent as 2018 in Syria, as reported to the United Nations.

Chlorine is used in modern chemical industrial processes relating to many industries. As the Greener Industry states, in the UK 1.6 million tonnes of chlorine is produced each year, with 60 per cent of all industrial processes relying on chlorine in some way.

The compounds and by-products from these processes can be very dangerous for the environment. It is the chlorine atoms in ozone-depleting chemicals that destroy ozone and are thereby responsible for the thinning of the ozone layer. It is the use of chlorine bleach in the pulp and paper industry that allows dioxins and other dangerous chemical by-products to get into the environment. The amounts are very tiny, but dioxins are some of the most dangerous chemicals ever known, and some sixty thousand times more poisonous than cyanide.

Dioxins are an accidental pollutant, created when using chlorine in the industrial process. Dioxins can be found in leaded petrol, some bleached white paper (including toilet tissue, cigarette papers and high quality advertising papers), weedkillers and herbicides. The most dangerous dioxin, 2,3,7,8-TCDD, is able to accumulate in the environment and in the body fat of birds, wildlife and humans (see 'Chemicals' on page 54).

Medical studies have shown that when people are exposed to too much chlorine they may suffer many side effects, including life-threatening illnesses in the lungs and heart. Chlorine can also cause skin irritations and blindness.

Chlorine causes lasting environmental damage and harm at low persistent levels and is considered damaging to the soil especially.

What can you do?

Chlorine is an interesting example of how our excessive greed for a chemical has had long-term effects on the ecosystem. Like the fossil fuel industry, the chlorine industry will soon have had its day. It is a corrosive industry, and cannot last in a zero-carbon, zero-toxin future. Our future selves will be amazed that we ever allowed it.

Scientific discoveries are finding new ways to clean bacteria from water, including the use of ultraviolet light. We can also expect new and exciting ways to treat viruses that currently require chlorine.

Boiling water has been a universal method of water cleaning for thousands of years. Many new forms of heating without using fossil fuels will be discovered and become the norm, making it a sustainable alternative to chlorination.

There are plenty of environmentally safe substitutes for chlorine-based cleaning products and household items. Use them! Avoid chlorinated bleaches and toilet cleaners: remember, everything you put into the water system has a long-term effect on our seas.

Using a charcoal carbon water filter at home – either in the form of a specialised jug or in charcoal pieces – will filter out the chlorine in your tap water. When they have exhausted their role as a drinking-water filter, these charcoal pieces have many end uses in your home: regulating moisture, keeping food fresh, halting mould growth or reducing odours.

Avoid whiter-than-white paper and soft tissue paper products that are bleached with chlorine.

SEE ALSO: Aerosols and the Ozone Layer, p. 25; Chemicals, p. 54; Hormones, p. 196; Pesticides, Herbicides and Agri-Chemicals, p. 244

CIGARETTES

The harmful effects of smoking on the body are well known. Smoking can cause lung cancer and is a factor in other cancers; it lowers the body's resistance to disease, such as heart disease. Smoking lowers fertility, causes many side effects and makes a person have long-term lung diseases even without the increased risk of cancer.

Pregnant women are strongly advised not to smoke: smoking increases the risk of long-term health problems for developing babies; it can also lead to premature birth and low birth weight. Smoking during and after pregnancy also increases the risk of sudden infant death syndrome.

There are also powerful environmental reasons for not smoking.

Cigarette smoke contains a variety of poisonous substances that pollute the air around the smoker and anyone else in the general area. A non-smoker can inhale the equivalent of two cigarettes a week if surrounded by people smoking. So-called second-hand smoking causes an extra 11,000 or more deaths in the UK each year. The poisons in the tobacco and cigarette paper include dioxins, nicotine, acetone, ammonia, toluene and arsenic. Carbon monoxide can be given off when a person smokes a large number of cigarettes in an enclosed space. Chemicals like chlorine and calcium carbonate are used to bleach cigarette paper. The cigarette, when lit, will release many other chemicals. The US Centres for Disease Control and Prevention (CDC) says there are 7,000 chemicals in tobacco smoke in a lethal chemical cocktail.

A cigarette is filled with rolled tobacco, grown as a lucrative cash crop. The crop has caused considerable deforestation, which in turn causes the loss of biodiversity and soil degradation. The environmental damage is devastating as region after region is transformed from a complex tropical ecosystem into a supersized tobacco farm. The chemicals used in growing despoil the waterways and rivers nearby, further weakening the local ecosystem. The amount of water needed to grow the crop is disproportionate to other crops; taking more from the ecosystem, often leading to drought and the drying of the local water table. Local people must then travel further to find clean drinking water.

The entire process, according to the World Health Organization report on the environmental impact of tobacco, causes considerable food insecurity, local drought and even hunger. The report is adamant as to its overall reach and impact on humanity: 'It also means that tobacco can no longer be categorised simply as a health threat – it is a threat to human development as a whole. This issue requires a whole of government and whole of society approach and engagement.'

And the damage doesn't end when you've smoked a cigarette. A discarded butt will take fourteen years to fully decompose and continues to leak its poisons long after the smoker has walked away. The most toxic form of all litter, cigarette ends, becomes food for small birds and sea creatures. Disposed of in the soil, in the streets, roads and gullies, a cigarette butt will end up in the rivers and in the oceans, where it can be eaten by fish and mammals. One discarded butt will contaminate about 500 litres of water.

The chemicals leach into the water and soil when a cigarette butt is casually thrown away. The effect is fatal for the water fleas known as daphnia, an essential part of the aquatic ecosystem. The tiny plastic particles in the filters take decades to break down. Incidentally, according to the World Health Organization, the plastic filter tips on cigarettes serve no use at all.

It is understood that around two thirds of the 5.6 trillion cigarettes with filters consumed per year are littered and end up in the oceans, rivers and earth. The World Health Organization said that: 'Tossing a cigarette butt on the ground has since become one of the most accepted forms of littering globally and borders on a social norm for many smokers.' What a tragic lack of common sense! What a sad and terribly consequential lack of awareness and reasonable social and ecological etiquette.

What can you do?

#TheSeaStartsHere is a creative arts campaign to help people realise that the route to the ocean is close by in a storm drain in your street. Artists from all over the world post their artwork concerning drains, pipes, sinkholes and plugs, often with an arrow and #TheSeaStartsHere.

Do not drop litter of any kind, especially cigarettes which are the most toxic. Pavement litter gets funnelled into the gullies as soon as it rains – the system is geared for it – and all that rain goes straight into the nearest river, and eventually to the sea or ocean, and tiny bits of plastic such as cigarette filters are hard to filter out.

Tell people who drop cigarettes of the damage they are causing. What's more, it is illegal to drop cigarettes as litter. There is no situation where it is acceptable to drop cigarettes or tobacco products and packaging on the ground.

When people put their cigarettes out in tubs and pots or in trees and bushes in streets and outside restaurants, they poison the plants. The poisons include nicotine and arsenic along with heavy metals. Every time it rains, the rainwater washes the poisons from the cigarette butts into the plants. Remember: they will take fourteen years to decompose. The plant will definitely die before the cigarette has decomposed.

If you smoke and want to give up, please seek help. Get support for your addiction before it gets you.

SEE ALSO: Chemicals, p. 54; Litter, Waste and Landfill, p. 218; Seas, Oceans and Rivers, p. 278

CLIMATE EMERGENCY

This is not an invention or a hoax. This is a real emergency.

Climate change is caused primarily by humans. Because of this we know we can change it. We already have the solutions. We already know the answers.

The Intergovernmental Panel on Climate Change, the scientific body of the United Nations, say these dangerous emissions come directly from burning fossil fuels and are the primary cause of climate warming. And they confirm it is human made.

It's not rocket science! Climatologists remind us that unprecedented rises in the global temperature and unprecedented drought and lack of rain encourage unpredictable and unprecedented fires. When ice melts, the sea rises. When extreme weather conditions worsen, floods and fires become commonplace.

The climate is warming mainly because the large quantities of fossil fuels we are burning produce pollution in the form of carbon dioxide emissions and heat. The pollutants and heat are trapped in the atmosphere as if there is a greenhouse surrounding the planet, making everything inside hotter.

Fossil fuels, mainly coal, gas and oil, are used to power our cars, fly planes, heat our homes and run our industries. These all create carbon dioxide and the amount in the environment has been increasing steadily since the end of the Industrial Revolution in the 1860s.

Modern agricultural practices can compound the problem. Chemical fertilisers for example produce nitrous oxide, another powerful gas contributing towards rising temperatures. Decomposition from animal waste and manure piling on agricultural land produces methane gas emissions. Methane emissions are found in smaller quantities in the atmosphere than carbon dioxide but each methane molecule is thought to be 84–104 times more powerful over twenty years than an equivalent carbon dioxide molecule in its ability to add to the trapped heat in the atmosphere. It is this trapped heat that is contributing to the temperature rising.

Cows emit methane when they fart and belch as well – a staggering 250–500 litres each day per cow. The vast number of livestock and agricultural farms on the planet contribute about 14.5 per cent of the total pollutants causing the climate to heat, as reported by the Food and Agriculture Organisation of the United Nations.

A new study says that seaweed could help cows to fart less if they eat it with their other food.

Leakages in the oil and gas industry from unstable and

corroding pipes make a significant contribution to the amount of methane in the atmosphere. As the Arctic ice melts more methane is also released from ancient stores in the soil and oceans beneath. These frozen molecules of methane are stored as natural gas in the permafrost and methane clathrate (a structure in which methane gas is trapped within ice). They are a naturally occurring form of methane but the melting ice is accelerating their release. The US Department of Energy warned us in 2008 that clathrate destabilisation in the Arctic would be the most likely scenario for an abrupt and catastrophic warming of the planet and life-threatening rising of sea levels.

The problem is exacerbated by the clearing of forests. Trees capture and keep carbon dioxide while supplying us with oxygen. Instead, as they are stripped and cleared and the land is handed over to agribusiness, the carbon they have been storing is released and this adds to the warming. The new agricultural business further escalates the problem with soy bean plantations, cattle ranching and other enterprises which also contribute to climate-related emissions.

This intensive human industrialisation has caused pollution levels and the overall temperature of the planet to rise, and at an alarming speed.

The climate crisis is directly leading to a global public health crisis. The World Health Organization estimates that at least 7 million people are dying each year from particles in air pollution from the burning of fossil fuels causing heart and lung conditions. They are direct victims of the climate emergency. Heatwaves also cause multiple medical problems including heat-related infections and extra deaths from dehydration and heat exhaustion. The changing temperatures have also increased insect infestations and the subsequent additional bites and ticks have increased related illnesses in humans. In the heat, food is more likely to spoil and there are increased numbers of food poisonings, and food rotting and spoiling early further exacerbates food insecurity and hunger. People living in areas of intense and sustained fires have suffered long-term lung and heart illness from the stifling polluted air. From fires to flooding, increased pressure on emergency health services and mental health services is common in communities suffering the complex effects of extreme weather conditions. Homelessness and exhaustion from environmental emergencies cause massive stress and can lead to PTSD.

Unless we do something to slow it down, the temperature average will rise 5 degrees by 2100.

A hotter planet has already caused many geographical changes and new problems. The melting polar ice caps are pushing the sea level higher and leading to more flooding in low-lying areas. Many countries are finding sea levels are rising much more quickly than they believed possible.

Until the early twentieth century sea levels had remained consistent for almost two thousand years, but in the 100 years since, they have risen nearly 20 centimetres. That is alarming. That seemingly small amount already causes havoc.

A sea-level rise of just 2 metres would be catastrophic to the world. In the UK, all low-lying areas would be flooded and areas like Cambridgeshire would be permanently under water. Parts of New York, London and Rio de Janeiro would also be under water. The world would find at least 187 million people without their homes and land, according to the IPCC modelling studies.

The East Antarctic Ice Sheet has enough ice to raise the global sea level by another 53 metres if it melts.

The signs are there, all around us. Extreme weather conditions have increased and worsened, according to the United Nations. If your home has been flooded because the local river burst its banks after extreme rainfall you will know the effects. Tens of thousands in the UK and millions of people across Europe and around the world have endured that in their homes and communities.

There are stronger typhoons and hurricanes now, more rain and heavier rain. This in turn causes landslides, flash flooding, and more damaging consequences to buildings and people and to the natural world.

If you have run away from a forest fire you will know climate change certainly is an emergency. In many countries, such as Australia, South America and the USA, drought and high temperatures make it all too easy for even a small spark to set the dry timber alight. Although wildfires on a small scale are natural and part of the forest cycle, the super-dry conditions have meant that the fires are bigger and there is a chain reaction that gets worse because of the climate crisis. The fires join up together, and they create their own weather systems, which cause lightning and even more fires. These wildfires destroy lives and communities. Climate change and pollution are turning extreme weather situations into very dangerous events, and causing apocalyptic conditions to develop very quickly. The future looks bleak if we decide to do nothing.

Humans are suffering already. In the first half of 2019, more than seven million people were displaced in emergency conditions by extreme weather events. Millions more will lose their homes if sea levels rise and once-viable land floods.

When drinking water is contaminated by salty sea water it becomes undrinkable. In low-lying countries and islands this has already become a reality. Imagine having to ship in water from another country because your own water supplies were no longer safe to drink?

As the temperatures rise in the atmosphere and the water, the seas and oceans become acidic because the oceans absorb carbon dioxide. This carbon dioxide reacts with the salty sea-water to create carbonic acid. Acidic sea water is dangerous for sea life. The coral reefs, home to 25 per cent of all marine species, are dying and the biodiversity of the ocean is severely affected.

Species are dying and becoming extinct on land and sea. This is the reality for the rest of the living species on earth. We are in earth's sixth Mass Species Extinction. A study by the University of Connecticut in the US says that 8 per cent of all animals will become extinct directly because of the climate emergency.

The coronavirus pandemic stay-at-home orders caused a massive shutdown of the world economic engine as people worked from home or closed their businesses entirely. For the first time in thirty years we all had visual proof of the effect of our industrialisation and travel on world air pollution levels.

The results were immediate and dramatic around the world, prompting several governments and the public to call for wholesale reductions in carbon dioxide levels and a switch to less polluting industries. Some governments denied carbon polluting businesses grants and loans to continue, investing instead in greener alternatives.

Greta Thunberg says, 'Thirty years of pep-talking didn't work. If it did emissions would have gone down already. We are currently burning 100 million barrels of oil a day. This has to stop. The rules have to be changed, everything needs to be changed.'

What can you do?

Your future life is in trouble. We are all in trouble. Don't listen to those who are still saying that environmentalists are extreme or scaremongers. Their fear and inaction should not get in the way of your determination. Everybody has to get on board. We all need to be 'green' now.

The solution to the climate crisis is a multifaceted approach. From protecting bumblebees and elephants to taking fewer car journeys and saving energy, there is a lot to do.

On the planet, right now, are all the people and inventions, the machines and the solutions that are needed. Many small countries have made their own trials; they have proved so much is possible. There are carbon-free economies. There are ways to grow vegetables without chemicals. There are countries that operate without burning fossil fuels. There are communities that have always survived and left no footprint whatsoever. All that knowledge is here, provable and successful.

Only when we work together, however, can we make a difference to the icebergs melting, to the methane that might be released or to the emergency of species extinction. Only when governments and businesses work together can we come up with solutions and make them happen. Those big companies that have known and continued in a 'business as usual' manner are now on trial. Those big companies that paid millions to shareholders and nothing for environmental mitigation or to repair damage are now on trial. We, as citizens, must tell them that this attitude is not good enough. We must demand more.

Who else is to speak for the species at risk of extinction if not you?

You are not alone in worrying about the climate emergency. Go and see your council representative, your member of parliament, your lawmakers. Tell them what you are feeling. Let them know what is worrying you. Ask them to tell the truth. They are the ones who must put emergency measures in place and turn us away from danger.

Use your consumer power, boycott companies that don't listen and contribute to polluting the planet. Tell them why.

Reduce your footprint. Adopt a one-planet lifestyle. Save energy. Don't waste things. Use green energy like electric cars and solar power. Change your energy supplier to one that funds renewables and energy efficiency.

This is crucial: know how much carbon your life is responsible for. The average person is using two or three planets' worth of resources in the UK. We only have one Planet Earth.

Stop shopping altogether unless it is for food or an emergency. See how long you can do this.

Stop buying plastic. Stop buying red meat and cheddar cheese: both use the most energy, water and resources to create their protein and therefore have a very high ecological footprint. Stop buying from agribusiness and start supporting local producers and farmers. Local shopping from artisans, bakers, cooks, farmers and growers reduces your footprint. Make buying local, organic and seasonal your priority in business.

Negotiate and barter, exchange and swap. Stop buying new things. Resist the emotional pull of new toys, new bright shiny electronic devices and plastic rubbish. Stop adding to the problem. Learn to love what you already have. Learn to expect surprises and gifts and stop the binge shopping. Give up on Amazon and online impulse shopping.

Stop buying things that have been flown and shipped around the world. Why do you need green beans from Peru (6,000 miles from the UK) when there is a seasonal glut in every allotment in the country? Or red apples from New Zealand (12,000 miles from the UK) when we have hundreds of British varieties? Buy local, eat especially in the season, and reduce the miles travelled and carbon footprint of the food industry.

Avoid household items that destroy the rainforests or encourage their destruction, further releasing carbon dioxide. Boycott products with palm oil in particular: get used to reading the ingredients labelling (see 'Tropical Rainforests' on page

304). Palm oil is in hundreds of everyday products especially cheap oils, butters, chocolates, prepackaged food and cake frosting.

Reduce your own travel and stop flying around the earth. Take the train or boat. Change your plans. This is an emergency. Don't fly at all unless you will personally pay the carbon price for it.

Plant more oak trees. Be part of the solution. A tree will absorb the carbon you have produced. Plant a thousand trees and offset your carbon for life, although they would take about forty years to work and you would need a large area of land! (We would need to plant 42 per cent of the planet with local trees to totally offset the carbon and stop climate change immediately, if we didn't do anything else.) The single best way to capture your own carbon footprint in the future is to plant a tree. The oak has the best carbon-absorbing qualities and is indigenous to Britain. Of course, tree planting will only work for our children in the future. It is, however, an important part of the solution; in the long run it will save and protect biodiversity.

Reduce your footprint further. The whole of this book is designed to help you to have knowledge about an issue and take personal action as well as being part of a growing number of aware people who won't hide from the difficult questions.

Open a bank account with a bank that cares about the planet. Banks like Triodos and the Co-operative Bank are ethical; they won't work with companies that fund oil and coal production, that are involved in wars or other businesses that don't meet their ethical standards. Don't just open an account with the first bank you come across, do the research. Don't put your money in a bank for the rest of your life in a business that won't align with your principles. Do the same for your electricity account and choose a provider that supplies electricity from renewable energy sources.

See how many of these actions you can accomplish. Work with friends to share ideas and find ways to achieve your goals.

Read the science. Get to know what you are talking about. There is plenty of information out there. The issue should not be about arguing the facts about the climate emergency any more, even if people tell you that. The scientists really do agree.

Ask your MP what they are doing. Have they called for emergency measures? How have they voted on environmental issues? Check their voting record before you meet them, on the site: TheyWorkForYou.com.

See more at #SixthMassExtinction #actnow #climateemergency #ourhouseisonfire #ClimateFact

Groups like Climate Action Network, Greenpeace, Friends of the Earth, XR (Extinction Rebellion) and the Transition Movement all campaign and work on the climate issue locally and internationally. The Climate Coalition is a nineteen-million-strong group representing 130 organisations in the UK working on climate issues. The UK Student Climate Network links together students under eighteen and organises hundreds of demonstrations. Fridays For Future organises the Friday School strikes.

Start a group in your school or college or local area.

Don't believe it is real? Simple. Read the IPCC reports on the planet, agreed by thousands of scientists. Read the science as Greta Thunberg tells us. Decide for yourself.

Watch Sir David Attenborough's BBC video *Climate Change: The Facts* on YouTube.

Remember, action can solve this. Action can take us from the brink to a new way of working with the earth. Don't think it isn't about you. It is about you. The solution includes you. We, all together, have to get on the programme to save the planet.

SEE ALSO: The Development of Humanity, p. 102; Economic and Environmental Business, p. 113; Ethical Living, p. 135; Fossil Fuels, p. 160; One Planet Living, p. 234; Species Extinction, p. 285

CONSCIOUS LIVING

When a person decides to be part of the green solution all sorts of things make sense. Your consciousness grows and you become acutely aware of the planet around you and what is necessary for its survival.

Planet Earth is naturally abundant, reproducing as is needed without lack. The planet is undergoing huge changes, however. The resources we are using are depleting our planet's natural ability to create abundance.

The toxins and poisons we create must go somewhere. When we strip the land of its trees and life we have to expect a consequence. When the soil is poor and overworked, the earth's life source is depleted.

When you decide to be conscious of what is happening around you it makes sense to change your lifestyle. You need to adapt, to make sure that your choices, your shopping, your holidays, job, and everything you do is done from a conscious and aware perspective.

You don't need to understand all the science to be conscious. Sometimes the right thing to do is obvious and clear, even common sense. Sometimes scientists have obvious and clear explanations for what used to be thought of as magical or impossible. In other areas, such as the chaotic, complex interrelationships all things have together, the complete picture is perhaps unknowable.

Being conscious opens the world of nature in your heart and

mind. Listen to nature. Spend time in nature in silence. Being connected and open-hearted to the voice of nature will enhance so many things in your life.

What can you do?

Witness who you are, and how you are. Be attentive to your own speech and words. Believe in yourself or change the script, the dynamic, and the situation.

Witness your footprint.

Forgive yourself and resolve to change, to do better. Carry out this change, this commitment.

SEE ALSO: Ethical Living, p. 135; Keeping Well and Safe, p. 213; One Planet Living, p. 234; Your Carbon Footprint, p. 322

CONSUMER POWER

Consumer power works! Sometimes long campaigns aren't necessary. Sometimes when a product on the market is seen to have a problem it can be solved without laws in parliament. Laws often take too long.

Over many years, consumer power has been an amazing and effective way for shoppers to show their displeasure and concern over issues. From animal testing to toxic ingredients, the things that upset us about industry are often challenged by consumers first, not lawmakers.

Consumer power creates additional pressure for companies to be socially and environmentally responsible. Consumers have the power to shift the thinking of companies, getting them to be more aware of environmental issues. Manufacturers, companies and supermarkets are very sensitive to the ideas of their customers and are conscious that they want to keep them happy.

Social media are also being used to drive change and new ideas in business. Young people are becoming especially vocal on issues that concern them and they are using social media to energise many others to follow suit.

In its simplest form, consumer power is achieved by you choosing one product over another when you are shopping. If you choose organic sweets, say, or make-up that is not only organic but is made by a company that is socially and ethically responsible, you are creating a bigger market for those products.

When you say no to a product that is filled with unnecessary or polluting chemicals, or is over-packaged, or is from a company that has a questionable social agenda, then you are also using your consumer power. You choose who you give your money to, and you also have the right to tell a company when you think they have got it wrong.

When you shop smart, you read the label, you want a product that is greener, that doesn't harm animals or people, that is made to last and doesn't waste energy.

What can you do?

Read the labels. Be aware of what is in the products you are buying.

Don't buy so much! Do you really need this product at all?

Boycotting a product doesn't necessarily mean going without, it can mean campaigning for something else as an alternative.

Be positive and support companies who produce good quality ecological alternatives. They can be small companies that are starting up and would be grateful for your support.

Text, email, write, tweet and use social media to say what

you want and to speak up for the earth and animals too. Tell companies how you want them to do business; you are paying for their service, after all, you are the buyer. Their service and products should reflect your needs and the limits of the planet at the same time.

Real consumer power is held when a citizen walks through life without any needs. A person in control of their purse is able to shop for what they want, not want what they are offered by companies with dubious or poor environmental standards.

SEE ALSO: Action, Protest and Empowerment, p. 16; Climate Emergency, p. 64; Consumption and Consumerism, below; Economic and Environmental Business, p. 113; Gifts, Presents and Greetings Cards, p. 167

CONSUMPTION AND CONSUMERISM

Do you have enough? Look in your cupboards right now. Do you have enough clothes or shoes and socks? If you are like the average person in Britain, Europe or the USA, you have more than enough. In fact, in Britain billions of pounds' worth of clothing lies in our drawers unworn, often still in its packaging, unopened!

Advertising has been very successful. It has hypnotised a society that feels a need, almost like a drug, for shopping. We crave the next new label or designer handbag, or the latest shoes, lipstick or computer game.

If you have felt the call to buy a product because of the name on the label rather than your actual need for it, you have succumbed to advertising and modern consumerism.

Our society has only recently developed our lives around having an emotional fill-up from shopping. A needy emotional and empty feeling can't be solved with shopping and eventually most people come to realise it. Binge shopping and addictive shopping, like any drug, can only give short-term feelings of satisfaction.

Addictive shopping can also be a serious disease. Some people find their addiction gets them into terrible debt, and hoarding issues arise in their homes. People with these extreme traits need counselling and supportive help, just like alcoholics and drug addicts do. They have often been found to have complex emotional issues with loss, with grief and even with eating disorders and other psychological disorders.

A consumer addiction for 'things' is literally destroying the planet. When you purchase any product you cause a chain reaction. From mining for raw materials to growing crops with pesticides for cash, every product has an effect. When you reduce your shopping you reduce this effect. It is a simple and logical result.

Companies that produce goods are worried about us not spending money and the impact that would have on their profits. They should concentrate on quality rather than quantity if they want to be truly sustainable and help us all live comfortable one-planet lives.

In the UK just ten companies control the mass production of the majority of processed food and drinks sold in supermarkets (see Oxfam's www.behindthebrands.org). They are Nestlé, who make, among other things, Perrier water, Kit-Kat bars and baby foods; PepsiCo, who make things like Pepsi and Lays crisps, Doritos and Tropicana; General Mills, who make Cheerios and Häagen-Dazs ice cream; Kellogg's, who make Pringles, Pop tarts and Special K cereal; Coca-Cola, who make Coke, Fanta, Sprite and various herbal and iced teas; Unilever, who make Hellmann's mayo; Mars, who make chocolate and Uncle Ben's Rice; Mondelez International (previously Kraft Foods), who make Oreos amongst other products; Danone, who produce bottled waters and Activia yogurt as well as nutrition products; and Associated British Foods, who produce Ovaltine, Twinings Tea and also own Primark clothes.

They all make thousands of products to sell all over the world but many rely on cheap labour abroad and at home. That means the workers often only get paid minimum wages, while the companies' vast profits go to the shareholders and the directors.

Many have terrible conditions for their workers. They disregard or undermine women's issues, such as maternity leave. The turnover of these ten companies is more than $1.1 billion every single day. Despite their profits and the length of time they have had this economic power, it's difficult to see progress in how they really empower or enhance the lives of their workers or the local communities that they rely on for workers.

The human body does not need to eat or buy any of the processed foods made by these brands to survive, to be healthy and to live. They are not essential to your diet in any way. Many have what we call 'empty calories', that is food filled with hidden sugars responsible for higher risks of obesity, type 2 diabetes, cancer and heart diseases.

This is how crazy our consumerism has become. We take a raw resource, like water, and add sugar, for example, to make a more expensive drink. The company amplifies its value by a huge margin so they can profit from the item and then they mass produce it.

Adverts hypnotise us into believing we really need that product to be happy so we go to a supermarket and buy it, remembering the nice feeling we had when we saw the advert. The production, disposal and health effects of plastic water bottles on the environment are multiple. The disposal of the product is a long-term problem. Our health and the environment are closely linked; here we see the detriment to the planet, to people and to our individual health, all linked, and all part of the same problem.

What can you do?

Reverse your life. When you are living a fulfilled and busy life there is no time for shopping.

Buy what is important and necessary, yes, but living more simply and sustainably means not falling prey to the allure and temptation to spend. Be content with who you are. Be at peace with who you are.

See how long you can go without shopping of any kind.

Grow your own vegetables. Upcycle your clothes. Swap with friends and neighbours for what you need.

SEE ALSO: Climate Emergency, p. 64; Consumer Power, p. 75; Economic and Environmental Business, p. 113; Gifts, Presents and Greetings Cards, p. 167; Litter, Waste and Landfill, p. 218

COSMETICS AND TOILETRIES

Of course we want to look good and smell good and there are hundreds of brands offering to help us do just that, promising glossy hair, dewy skin and that white 'Hollywood' smile.

In July 2019 the global beauty industry was valued at $532 billion, according to retail analytics firm Edited. The industry has also been boosted by the male grooming market which has grown in the past few years: according to Statista this is expected to be worth more than $80 billion by 2024.

However, the products we use every day are not only costing us a lot of money, they are also creating pollution in the environment, curtailing animals' lives and our health.

The products we use may be full of hormones and chemicals, crushed insects, animal urine and other parts of animals. Clinicians are now treating a growing number of people for a range of hormonal problems, allergic reactions and even psychological complications from cosmetic chemical overload. (See 'Hormones' on page 196)

When we continuously use products containing industrial chemicals on our skin we are lowering our resistance to fight disease, exposing ourselves to possible cumulative negative reactions. The toxins the products contain become part of us; our bodies absorb the chemicals.

Additionally, cosmetic packaging is mostly made of mixed materials which are hard to recycle and therefore end up in landfill, adding to the tonnes of non-recyclable waste.

Below is a list of the most damaging ingredients commonly found in beauty products and the effects they are having on the environment and the human endocrine system.

Alcohol

Alcohol is commonly found in hand and body wash, toners, perfumes, mascara and eyeliner, make-up remover and nail-varnish remover. SD alcohol, denatured alcohol and isopropyl alcohol are volatile alcohols which give products quick-drying properties and are commonly found in products for oily skin and acne. However, their drying properties are damaging to your skin's natural barrier, as your skin will over-compensate for the oils lost. Alcohol can also cause redness and irritation and weakens the skin in a way that is irreversible.

There are some types of fatty alcohols such as cetyl alcohol, stearyl alcohol and cetearyl alcohol which in moderation are non-irritating and can be beneficial to skin and the quality of products.

Fragrances

Most conventional cosmetics or toiletries contain fragrances, which might include a mixture of ingredients, to make the product – perfumes and aftershaves, soaps, hand and body washes, shampoo and conditioner, fake tan lotions, sprays – smell alluring. According to a report 'Fragrance in the workplace is the new second-hand smoke' by researchers at the University of Maryland, USA, of the 4,000 chemicals currently being used to create our favourite scents, manufacturers can use from 50 to 300 separate ingredients in order to make a fragrance. If it's labelled as 'fragrance' or 'perfume', most ingredients used will be synthetic. You are unlikely to know the full ingredients.

On the whole, people are completely unaware of the effects their fragranced products will have on them and on others around them. Symptoms from overuse or allergic reactions to synthetic 'fragrances' can include headaches, skin irritations and skin discolouration. Some people report psychological reactions from exposure to fragranced products. The same chemical fragrances can be found in many household products too.

Mineral Oil

Mineral oil comes from the earth and is a natural ingredient; however, concerns are raised by the process by which it is extracted, processed and refined. Mineral oil is petroleum, a by-product of the petrochemical industry, making it non-renewable. Mineral oil can be listed as an ingredient under different names such as paraffin oil, white oil, liquid paraffin and more.

It is commonly used in lotions, creams, lip products and

make-up removers due to its ability to form a protective layer over the wearer's skin to lock in moisture. Although cosmetic-grade mineral oil is highly refined, we must take into account that it is still distilled from petroleum, and untreated mineral oil is known to be a carcinogen.

Petroleum jelly and mineral oils can also have the effect of skin drying and the promotion of acne and other skin disorders such as nappy rash. Some have the effect of causing premature skin ageing.

Plastics and packaging

Disposable, single-use products such as cotton buds, cotton pads and wet wipes cause environmental problems, especially in our waterways.

Most wet wipes contain plastic resins like polyester and polypropylene and they can never fully biodegrade: they simply break into smaller and smaller pieces, releasing micro-plastic into the environment. (Microplastics are any kind of tiny plastic particles smaller than 5mm.)

Cotton buds and cotton pads are generally made from conventional cotton; to find out more about the environmental damage associated with that see 'Cotton' on page 88.

Don't dispose of anything apart from human waste in the toilet. Wipes, cotton buds and cotton pads must always be disposed of in general waste bins. According to Friends of the Earth, in 2017 wet wipes made up more than 90 per cent of the material causing sewer blockages. They are a major cause of blockages in the notorious 'fatbergs' found in London's sewage system.

Microbeads are a type of manufactured microplastic which is used on the skin as a scrub or exfoliator. In response to the increasing demand for beauty products to be plastic-free, in

2018 the import or manufacture of toiletries containing microbeads was banned in the UK.

There has been a growing trend to introduce refill stations for beauty products, and reusable products for removing make-up and cleansing, in an effort to reduce the amount of single-use products and packaging produced. Refill stations encourage customers to take their existing cosmetics containers into stores and refill them without the need to buy new bottles. The Body Shop was one of the first to offer this service over forty years ago, and the Beauty Kitchen and UpCircle are among those retailers encouraging refilling today.

Reuse and plastic-free is becoming increasingly popular with many ethical companies producing non-plastic-stemmed organic cotton buds and organic cotton and hemp or bamboo reusable, washable make-up pads and cloths. These are a great way to cut down on waste and save money in the long run.

Preservatives

Synthetic preservatives have been used in cosmetic and toiletry products since the 1920s. Preservatives are added to products to prevent the growth of harmful mould and bacteria. However, scientific studies have shown that they disrupt hormones in the body and harm fertility and reproductive organs, affect birth outcomes, and increase the risk of cancer.

A common preservative used in cosmetics and toiletries are parabens. They are typically added to soaps and body washes; cleaners, toners and moisturisers; foundations and primers; blushers and eyeshadows; lipsticks; fragrances; make-up removers and wipes, and toothpastes. Studies have shown that parabens can increase the user's probability of developing breast cancer cells and are potential endocrine disruptors due to their ability to mimic oestrogen.

Another group of preservatives used in personal care products are triclosan and triclocarban. Commonly found in antibacterial handwashes, dental products and antiperspirants and deodorants, studies have shown that triclosan and triclocarban are both dangerous and in fact unnecessary. The US Food and Drug Administration banned their use in household liquid soaps in 2016 but an EU scientific body thought they were safe to use. 'Consumers may think antibacterial washes are more effective at preventing the spread of germs, but we have no scientific evidence that they are any better than plain soap and water,' said Janet Woodcock, MD, director of the FDA's Center for Drug Evaluation and Research (CDER) in 2016. 'In fact, some data suggests that antibacterial ingredients may do more harm than good over the longterm.'

Phenoxyethanol is used as a preservative in beauty products to limit bacterial growth and as a stabiliser in fragrances and perfumes. It has been known to cause reactions from eczema to life-threatening allergic responses. Infants especially should not be exposed to phenoxyethanol. Phenoxyethanol can be found in creams and lotions, shampoo and conditioner, soaps and body washes, make-up, lip products, sunscreen, hair-styling products, nail polish, baby wipes, shaving cream, deodorant, toothpaste, fragrance, hair removal waxes, hand sanitiser and ultrasound gel.

Retinol

Retinol and retinol compounds are a micronutrient vitamin A which can be harmful to health and which are added to cosmetic products in two forms: retinoic acid and retinyl palmitate. Commonly found in anti-ageing creams and lotions, moisturisers and foundation, this ingredient should be avoided. Studies have shown retinol and retinol compounds can cause cancer and developmental and reproductive toxicity.

Sulphates

The cleaning part of your personal-care items is likely to be made of detergents in the form of sulphates. They are known as: SLS (sodium lauryl sulphate), SLES (sodium laureth sulphate) and ALS (ammonium lauryl sulphate). Their main use is to cut through dirt and grease, but unfortunately they can also strip away the skin's natural oils, causing or exacerbating skin problems. SLS and SLES can be derived from petroleum oil, coconut oil or palm oil.

Studies show both SLS and SLES can cause skin irritation, eye damage, diarrhoea, respiratory damage, cancer and death in laboratory animals. Cosmetics and toiletries that commonly use sulphates include soap bars, hand or body wash, make-up removers and toothpastes.

Talc

Talcum powder is a natural, fibrous clay mineral mined from the earth. It is known as cosmetic talc when used in cosmetic and personal care products such as baby powder, lipstick, mascara, face powder, blusher, eye shadow, foundation and child-friendly make-up. Some talc products have been found to contain asbestos deposits, a known carcinogen. Exposure to talc can also expose the user to irritations and organ system toxicity.

Asbestos-contaminated talc is restricted in the EU, and the talc available should be asbestos-free. However, as documented in *Beauty Laid Bare*, by the BBC, investigations into powdered make-up products claiming to be asbestos-free, have found traces of asbestos. Although there is doubt whether pure talc is harmful, asbestos is undoubtedly dangerous. A recent investigation in 2018 by Reuters found that many talc mines contain asbestos and are the source of contamination.

Whitening Agents

Titanium dioxide (TiO2), found in skin care and make-up products and widely used in toothpaste, works as a UV filter or whitening agent. It is found in sunscreens and pressed and loose powdered make-up products such as blusher, eyeshadow and foundation. Studies have shown that TiO2 in creams and lotions poses a low risk to the wearer; however, according to the International Agency for Research on Cancer, if inhaled in its powdered form it is considered a carcinogen.

What can you do?

Get familiar with toxic cosmetic ingredients. The Campaign for Safe Cosmetics by the Breast Cancer Prevention Partners provides extensive details of ingredients to avoid in their 'Red List'.

Use your purchasing power to good effect. Read the labels and check the ingredients. Download the Think Dirty app.

Buy cruelty-free cosmetics and toiletries, and look out for cruelty-free logos on products. If consumers demand products made only of natural ingredients, which are not tested on animals and are packaged sustainably, it encourages brands and manufacturers to evaluate the environmental impact of their products and processes.

Shop for products labelled 'fragrance-free', or why not try making your own beauty products? A quick online search will bring up a number of tutorials for 'DIY' creams and lotions.

How about looking for low-waste products sold by companies who market their cosmetics and toiletries in refillable containers, such as All Earth Mineral Cosmetics and Bower Collective?

Look for zero-waste, organic and natural brands.

Try to cut down gradually on the amount of toiletries you are buying and using. Only buy what you need, when you need it, and always look for sustainable products with minimal packaging.

There are many online blogs demonstrating how you can make your own reusable wet wipes. And remember to never flush wet wipes down the toilet, even if they claim to be 'flushable'. Alternatives to these removers are soft organic cotton flannels and reusable make-up rounds. An alternative to disposable cotton for removing nail varnish is to use squares of dark felt fabric. These should be washed in cold water after use.

Look for organic baby powder from retailers such Neal's Yard and alternatives to talc such as cornflour or oat flour.

Think about what makes someone beautiful. We believe that real beauty is something a person finds in themselves. Beauty within manifests itself as kindness, compassion and true friendship.

SEE ALSO: Aerosols and the Ozone Layer, p. 25; Allergies, p. 33; Animals and Insects in the Cosmetics Industry, p. 35; Animal Testing, p. 45; Chemicals, p. 54; Cotton, p. 88; Deodorants and Antiperspirants, p. 95; Hair Products and Hair Accessories, p. 183; Hormones, p. 196

COTTON

Cotton has been in use for thousands of years all over the world. In Mexico, scientists have discovered traces of cotton balls and pieces of cotton cloth in caves that could be at least 7,000 years old. Analysis of that cotton told scientists that not much has changed in the way cotton is produced and harvested today.

By 1500, cotton was known and used as a crop worldwide. In 1730s England, cotton was first spun by machinery. The Industrial Revolution in England and America paved the way for cotton to become the important crop that it still is today.

Cotton has many uses; it is used in fashion, homeware, beauty, feminine hygiene products and much more. Almost every part of the cotton plant can be used; a sustainable side of cotton harvesting is the lack of wastage.

Unfortunately, however, the adverse environmental impact of cotton production really outweighs the positives. A significantly larger area of land and greater quantity of water is needed to grow organic and non-organic cotton than other textile fibres. For example, in order to produce 1kg of cotton, the equivalent to a single T-shirt and a pair of jeans, more than 20,000 litres of water are used, as reported by the World Wide Fund for Nature (WWF). In comparison however, more sustainable textiles such as organic hemp require much less water. For example, for the comparable amount of fibre in weight, hemp requires 300–500 litres, according to hemp developers and producers SeFF.

You may think of cotton as a natural product, but conventional (non-organic) cotton can be full of chemical pesticides. According to The World Counts foundation, cotton accounts for 16 per cent of global insecticide use. The use of pesticides in cotton harvesting is having detrimental health effects on the farmers who grow the crop. In the US more than 10,000 farmers are thought to die every year from cancers directly related to these chemicals according to industry sources. What's more, carcinogenic dust from the farms is carried by the wind into nearby residential areas causing and exacerbating respiratory diseases.

In 2014 cotton production was directly linked to the environmental devastation of one of the world's largest lakes, the Aral Sea, located between Kazakhstan and Uzbekistan, which has now completely dried out. The bottom of the lake, now exposed, has released salts and pesticides into the atmosphere, poisoning both farmland and people. The effect of this water-guzzling industry has meant long-term devastation of local communities and wildlife in the ecosystem.

The Cotton Campaign brings together business organisations who work together for human rights within the cotton production industry. For example, they campaign against the use of forced labour in cotton harvesting, such as occurs in Turkmenistan and Uzbekistan and many other cotton-producing countries.

With a view to establishing a more sustainable cotton industry, the Fairtrade movement works with cotton farmers in developing countries who make their living from cotton harvesting. These cotton farmers are constantly facing challenges, such as the impact of climate change and the artificially low prices of cotton caused by rich countries who are harvesting cotton at abnormally low prices and with whom the farmers simply cannot compete economically.

Organic cotton is usually certified by GOTS (The Global Organic Textile Standard). GOTS are responsible for monitoring the worldwide standard requirements for organically grown cotton. They ensure the organic status of the farming of the crop, and monitor the environmental and social responsibility of manufacturing in order to provide an ethical product and service to the end consumer: you.

Organic cotton still requires a lot of water to harvest; it requires more farming land and is more labour intensive. It is, however, still the best option when considering products made from cotton and providing it is used to its fullest lifecycle potential it is an important addition to sustainable textile production.

Cotton production should be about making a special, rare commodity, not something we expect every day. It should be used and reused when possible. Only one generation ago our grandmothers handed down the best embroidered cotton tablecloths or linen fabrics to their children. It was valued and looked after.

Cotton is such a core fabric, universally used; we touch it every day and take it for granted. Cotton makes every household complete. We have cotton clothes, kitchen tea-towels, bedding, rugs, mats and curtains. Imagine if there was no cotton underwear.

What can you do?

Cotton recycling and reuse will be a massive industry in the future.

When buying anything made from cotton, always opt for organic, fair-trade, unbleached cotton, which does not use unnecessary chemicals; look out for the GOTS logo. There are now many retailers who are producing collections using organic cotton, and even some brands that only use organic cotton and other natural fibres without the use of chemicals. Ask retailers where they are sourcing their cotton from.

Support the Cotton Campaign and the Clean Clothes Campaign.

Avoid cotton and synthetic mixtures, as it is more likely that bleaches and chemicals will have been used to produce them and they are harder to recycle.

Always recycle your cotton. Textile recycling reduces the amount of raw materials and water needed overall and is a future fashion trend.

SEE ALSO: Agriculture, p. 27; Fashion and Clothing, p. 139; Sustainable Textiles, p. 288

CYCLING

Cycling is the most environmentally-friendly and energy-efficient mode of transport in the world today. Since the invention of the bicycle in 1817, and the pedal version in 1840, billions of people have used them to get around, to keep fit and to carry great weights with ease. Bicycles have been supporting people without creating air pollution. They are able to carry about 300 pounds or 136 kilos in weight.

The World Economic Forum recently declared that the best cities for a cyclist were: Utrecht in the Netherlands, Münster in Germany, Antwerp in Belgium and Copenhagen in Denmark. They decided that the factors for the winning cities included bike-sharing schemes, infrastructure, special schemes and programmes for cyclists, and crime and safety issues.

A bike needs no energy other than your own legs. It is an empowering and simple form of transport for millions of women and men in developing countries. One of the key indicators of a family getting out of poverty in a developing country is when the family get a bike and are able to move more quickly and carry greater loads.

If you live in a city you will often find cycling gets you to your destination far more quickly than a car. You will see cyclists weaving among the stationary traffic and whizzing past everyone else. Road safety is obviously an issue, but the number of cyclists who have died on the roads has reduced over many years.

In 1987 in Britain more than 280 people died on the roads

on their bikes according to government Department of Transport statistics. More understanding of the importance of cycling helmets has helped reduce fatal head injuries. In 2018 99 cyclists died on the roads. However, there were still 3,707 serious injuries, and, in all, more than 13,744 accidents where cyclists were involved but did not have a serious injury.

In the same time period in the UK during 2018 some 1,770 car drivers were killed on the roads; 26,610 people were killed or seriously injured, with 165,100 people involved overall as casualties of all severities. That number has stayed almost the same since 2010. Our roads are inherently dangerous.

Cycling is less dangerous than driving. It is healthier, but has obvious restrictions in terms of distance and weather. It is a perfect carbon-free transport method, especially for the 40 per cent of journeys that are less than 2km each day.

What can you do?

See the world on two wheels. It is an exciting and simple way to take a holiday or explore a city.

Learn to cycle and take care of yourself by learning the Highway Code to know and understand the laws of the road.

Always wear a helmet and take note of the potentially crazy driving of other road-users. Drivers are not always easy to second guess on the road and they do strange things when they are under pressure or lost.

Get off and walk for a bit if you are worried. Wear safety coats and reflective gear when cycling in dangerous conditions or in the dark. Good quality rainproof clothing will keep you on the road during a shower, too.

You should feel really pleased with yourself if you manage to cycle a lot. It is quite simply an excellent way to travel. Cycling 10km each way to work would save the equivalent emissions of 1,500kg of carbon dioxide in a year (see 'Your Carbon Footprint' for a clear explanation, p. 322).

The health benefits are brilliant, too. Your heart and lungs will thank you.

Contact Sustrans UK for information about the National Cycle Network, and Cycling UK and the London Cycling Campaign for cycling information,

SEE ALSO: Transport, p. 294; Walking, p. 315

D

DEODORANTS AND ANTIPERSPIRANTS

It is completely natural to have body odour, to sweat, even to smell.

It's also normal to care about the way our bodies smell. In fact, the practice of making our bodies smell nice has been around for thousands of years: the Egyptians used pure perfumes from flowers grown beside the Nile river.

The first invention of deodorant in a stick form dates back to 1888, with the first antiperspirant created in 1903. That first antiperspirant was so acidic that it could eat through clothing! Not that much has changed; the sticky substance still has a reputation for ruining the wearer's clothing by leaving unwanted stains on garments.

Today, we buy millions of deodorants and antiperspirants every year to help mask the smell of our own bodies. However, perspiring is a natural process. Our bodies need to get rid of all sorts of chemicals and toxins and salts that build up in the bloodstream. Washing every day allows these natural processes to defend your immunity. Exercising every day enables the toxins to exit the bloodstream safely through sweat. You wash it off. Simple.

When you use a deodorant, and especially an antiperspirant, you try to stop this natural process. Conventional deodorants work by using an artificial fragrance to mask or mitigate the smell created when we sweat.

Antiperspirants use aluminium chlorohydrate salts to block the pores in your underarms, thereby reducing or eliminating sweating. Aluminium has been found to cause DNA mutation, which can trigger cancers. Aluminium has also been found to cause issues for people with weakened kidney function.

Deodorants and antiperspirants can cause allergies; they can also irritate or even harden the wearer's skin.

The ingredients and their packaging, often in aerosols and plastic, can be an environmental hazard too. The sourcing process of the ingredients found in deodorants and antiperspirants can often be energy and resource intensive.

What can you do?

Some people are more prone to perspiration than others. It is natural to perspire. It is unnatural to try to stop yourself perspiring. Wash daily if you can; use a flannel, which you can rinse out at a sink at work or school, to keep you fresh.

Try not to use deodorant or antiperspirant at all for a week or a month and let your body readjust to your natural smell. Try to notice any differences you may feel.

Use natural, cruelty-free products if you need to, such as crystal salts. Ensure all materials and ingredients are responsibly sourced.

Avoid using antiperspirants in aerosol packaging. Transition to a natural deodorant like salt or aloe, or try simple ingredients like bicarbonate of soda.

Look for deodorants in a natural packaging such as card or cork. You can source these from Earth Conscious, or find refillable options offered by brands like Fussy.

Consider making your own deodorant to suit your skin type. There are many recipes online that can help.

SEE ALSO: Animals and Insects in the Cosmetics Industry, p. 35; Animal Testing, p. 45; Cosmetics and Toiletries, p. 80

DESERTS

The largest desert in the world is the Antarctic Desert, followed very closely by the Arctic Desert; both cover more than 5,000,000 square miles. The Sahara Desert is next at about 3,000,000 square miles.

Deserts can be spectacular, beautiful and hold extraordinary wildlife, albeit in small numbers.

Did you know that the Sahara Desert was once an incredible forest? Human beings were responsible for destroying the forest through desertification: they stripped all the living trees from the area and the land dried up and eventually turned to sand. The weather and human interaction changed the area from a lush tropical forest – producing an abundant food supply in Roman times – into the sand that everyone knows today.

Deserts are harsh, dangerous and inhospitable places. Climate changes, including rising temperatures, mean more wildfires and drying arid land. Waterholes have been drying up in desert areas, making them even more inhospitable. Plant and animal species that survive have adapted to such a harsh environment.

More land becomes desert every year. Poor farming techniques and land management practices have stripped the land dry in many countries in the Pacific, the Caribbean and Africa.

Natural habitats and biodiversity are at risk across the globe. When the few inches of precious topsoil is lost, it is generally irreplaceable, along with the local species and ecosystem. Eventually the land becomes a desert.

Deserts are vulnerable and fragile ecosystems. The delicate soil takes a long time to recover if it has been disturbed or changed.

What can you do?

Support The Young People's Trust for the Environment and WWF who are working to protect desert ecosystems.

Action Against Desertification is a group of countries working together within the UN to beat desertification with positive projects. Already they have reached and involved 500,000 people and planted 4.3 million tree seedlings.

Support organic food production.

Make sure you are enhancing the soil that you are responsible for in your own garden or allotment. Follow the Permaculture principles of soil management.

SEE ALSO: Climate Emergency, p. 64; The Development of Humanity, p. 102; Habitats and Biodiversity, p. 180; Soil, p. 283

DETERGENTS, CLEANERS AND YOUR KITCHEN

Did you know that the average person will spend two years of their life cleaning and washing up! The everyday detergents we all use to keep our environments clean and germ-free include washing powder, bath and kitchen cleaners and washing-up liquids.

We would never imagine that in fact we could be using dirty, toxic ingredients to achieve all this cleanliness.

Cheap, conventional detergents can contain nasty chemicals such as 1,4-Dioxane, which can cause skin irritations. It is found in almost all solvents and industrial cleaners and is considered to be a carcinogen by the US Environmental Protection Agency.

When manufacturers create an instantly recognisable detergent smell, they do that by adding a fragrance. However, just as in the production of cosmetics (p. 80), manufacturers can use anything from 50 to 300 separate ingredients to make our cleaning products smell like fresh fruits or even fresh air. Most of the ingredients used will be synthetic. Many people are completely unaware that the fragrances they use to make their environments smell nice and appear clean are in fact having a toxic effect on their bodies. Symptoms from overuse or allergic reactions to synthetic 'fragrances' can include headaches, skin irritations, skin discolouration and psychological reactions.

Surfactants (surface active agents) and bleaches are also found in detergents. Surfactants work to separate any fats and oils from the surfaces being cleaned. They have the same effect on the skin as fragrances, with drying and skin irritations recorded.

Chlorine bleach is highly toxic and very strong. It has a distinctive smell. When bleach enters the water supply, it can kill the good bacteria the ecosystem needs, for instance to break down enzymes.

Laundry liquids and powders, targeted to keep your whites whiter than white, will often contain optical brighteners, and enzymes which work to break up protein stains such as chocolate and egg. Both have been known to cause allergies and are dangerous to aquatic life especially.

Direct contact with detergents can cause skin irritations. Simply inhaling the chemicals is thought to contribute to cancer and birth defects. Bleach is especially dangerous. A study by Harvard University in 2017 found that using bleach once a week could increase the chance of developing COPD – chronic obstructive pulmonary disease – by nearly one third.

Some cleaning products have been found to contain a substance called NTA (nitro-tri-acetic acid). This substance combines with heavy metals to form soluble compounds. These heavy metals, which would normally remain at the bottom of lakes and rivers, are then reintroduced into the water supply, even ending up in our drinking water.

The products we use have to go somewhere, they don't just stay on the surfaces where we spray them. Eventually, all detergents will wash out into the water supply through the drains, and many chemical detergents will cause long-term environmental damage and pollution. If they are toxic to humans they will most definitely be toxic to wildlife. The environmental effects are evident.

Studies have shown that when chemicals react with the nitrates in our water supply from fertilisers, it promotes the growth of algae, found in rivers and lakes. Algae deprive fish of oxygen, resulting in their death. Many lakes have been damaged from green algae blooms taking over and the blooms are highly poisonous to humans, dogs and wildlife.

Phosphorus was used in detergents in the 1970s and 1980s, which led to the eutrophication of water: the water became enriched and full of algal blooms. The European Union and other industrial countries moved to ban its use in industrial processes and detergents especially. Waste water treatment plants were also improved.

As new chemicals are invented to bleach and whiten our clothes, concern grows again for their long-term effect on natural waterways.

Clothes that are made of nylon, acrylic, elastane and polyester mixes, all plastic-based, are also causing problems in the environment. Every time they are washed, especially in a washing machine, they release minute pieces of microplastics. These are smaller than a millimetre and make their way into the water supply through the sewage when washing takes place. They too bio-accumulate in our bodies.

Thanks to growing concern from consumers, more environmentally-friendly cleaning products are now available, and can be purchased in most supermarkets, wholefood shops and in refill centres.

What can you do?

Try to buy products which are eco-friendly, biodegradable, cruelty-free and vegan. Look for plant-based products (not oil based).

Don't put bleach down the sink or toilet and avoid where possible.

Watch out for detergents in other products. Did you know, for example, that shampoo and bubble bath both contain detergents?

Do you have a local zero-waste store where you live? See what environmentally-friendly household detergents they have available in refills.

Eco-balls, stones, soap nuts and a range of alternative reusable washing solutions are offered by various companies. Check your local wholefood store for the latest inventions.

Consider making your own detergents to suit your needs. There are many online tutorials that can help. Ordinary table salt makes a good scouring powder and white distilled vinegar mixed half and half with water is good for cleaning glass, tiles and even toilets.

People are finding that washing their clothes in a machine without using any detergents at all seems to be as good as cleaning clothes with the products. This is a revelation to many people and could instantly save a household hundreds of pounds a year. Try it for yourself.

SEE ALSO: Chemicals, p. 54; Chlorine, p. 58; Cosmetics and Toiletries, p. 80; Fashion and Clothing, p. 139

THE DEVELOPMENT OF HUMANITY

> 'Do not be dismayed by the brokenness of the world. All things break. And all things can be mended. Not with time, as they say, but with intention. So go. Love intentionally, extravagantly, unconditionally. The broken world waits in darkness for the light that is you.'
>
> L. R. Knost

People who are facing environmental disaster, stress and emergency are in need of help. Our response as humans should be to help with compassion and strength, resolution and kindness.

When our response is *not* based on human compassion, hope is lost.

We have seen insipid, fearful politicians. We have witnessed companies who destroy the planet be exposed, revealing their greed and avarice. They are behaving badly, and no matter how much politicians and industrialists bluster and accuse us of scaremongering and lies, the truth has been emerging for decades. Many leaders have been putting off tangible action for ten, twenty, even thirty years already.

Time is up on this bad behaviour.

The earth is in an emergency, and citizens, world governments and businesses need to work together to start to reverse the problem.

Environmental education, including climate emergency education, should be mandatory; everyone needs to learn. We need to learn more about an alternative way of living and why it is necessary for our survival. We need to stop the waste, to preserve what we have.

Behavioural changes are hard: we resist most things that frighten us. We don't want to know, the future is too awful to contemplate. We rationalise that we might survive without doing anything about it. We deny its importance or imminence. This is called cognitive dissonance, and millions of people are in this stage of their denial right now.

Naturally people want to keep what they have. A survey in 2019 found only 7 per cent of people were willing to give up their car for the sake of the planet.

Negative information makes us panic and feel bad. It's that simple. Our brains won't allow us to listen to bad news and try to shut down. It is difficult to get people to understand the risks and dangers and then to act.

Governments and policymakers need to do something effective for whole countries. Governments can shift public opinion and help people become more positive and hopeful when huge environmental changes are underway. We can look to the coronavirus examples. New laws could force companies and industries to change their ways, to be cleaner, or to use less dangerous chemicals.

Sustainable and greener businesses are also better for economic growth. We need political policies that insist on a world that is carbon neutral and ethical.

To be ethical, to find love for our planet in each other despite our differences, in opinion and culture, in politics or religion, is a powerful way to be a force for change.

We need moral leadership from those in charge. We need leaders with strength and integrity, leaders who can stand up to those international companies who ignore their environmental responsibilities.

There are countries that already pay fair wages, benefits and for healthcare when needed for all their citizens. There are companies who prioritise the environment and have cut their carbon drastically, even reducing their waste to zero.

Peace, equality and social justice are attainable goals if money, education, training, leadership and policy are set on paths of sustainability. Sustainability is an achievable goal, and one that people can get behind and be proud to be a part of.

Dr Martin Luther King Jnr said: 'We need leaders, not in love with money, but in Love with justice. Not in love with publicity, but in Love with humanity.'

We can extend that cry for justice and love for humanity to our planet's wellbeing.

Surely then the change to come must be about the heart and love? When we love people, animals and nature, our instinct is to make the earth cleaner, better, not to destroy her.

The further away a person is from remembering love, the worse their choices for the planet and for the future of humanity.

It is not a weakness to understand this love, it is a sign of emotional intelligence. The planet needs us to hold these thoughts and visions of peace, sustainability and harmony with nature more than ever.

What can you do?

What are your personal ethics? Do you care who makes your food and drinks? What do you knowingly participate in that destroys the planet?

What did you learn from the world's response to Covid-19?

How will love be part of this impulse for change?

Reach down and touch the earth. Show it gratitude. Everything you eat comes from the earth. Everything you wear and your surroundings come from the earth.

Try to be a world citizen who honours the earth for her abundance.

SEE ALSO: Climate Emergency, p. 64; Conscious Living, p. 74; Consumption and Consumerism, p. 77; The Development of Humanity, p. 102; Litter, Waste and Landfill, p. 218; One Planet Living, p. 234

DISPOSABLE PRODUCTS

Sir David Attenborough, the extraordinary filmmaker and environmentalist, said it best:

'The best motto to think about is not to waste things. Don't waste electricity, don't waste paper. Don't waste food. Live the way you want to live but just don't waste. Look after the natural world, and the animals in it, and the plants in it too. This is their planet as well as ours. Don't waste them.'

Disposability is a modern invention. Many people from indigenous and remote tribes had one bowl and one cup for their whole lives. People used products until they were utterly worn out and completely unusable. The idea of disposability never entered people's minds. When the first industrial potteries were invented primarily in the Roman era, small clay cups and bowls were made, and many have been found in archaeological digs, broken but essentially used continually before disposal.

We use disposable products every day without a second thought. They can be single-use and limited multi-use. Packaging, cutlery, cable ties, disposable razors, nappies and wet wipes are just some examples of everyday disposable items we are using. They are rapidly filling landfills. When we throw something away, it has to go somewhere, nothing just goes away.

We have become accustomed these days to using resources as if there is an unlimited stock, a supply of everything, all so easy to find and create. It is not true. There is only one planet and a finite amount of resources on it.

When consumers get used to disposability, nothing has long-term value any more. The problem is that the planet can't do it. If you use one cup at home over and over, washing and rinsing it, then it remains reusable. If you use a disposable plastic cup and throw it away after one use, that cup, or that plastic drinks bottle, won't be reusable but will still be around for about another 450 years!

Most plastic and one-use products are thrown away immediately. So many things in our society now are disposable.

The coloured caps on some soft drinks resemble the middle part of a sea creature that birds eat. When they are disposed of without care, some end up in the ocean. Seabirds then mistake them for their normal food and they eat them. They go back to their nests and feed their own tiny baby birds plastic! We know they are dying with large amounts of plastic in their stomachs. Thousands of birds have been found dead in remote island regions and some with more than 250 individual pieces of plastic in their stomachs.

The production of a single disposable paper coffee cup requires 33g of wood, 4g of petrol, 1.8g of various chemicals, 4 litres of clean water and 650 British thermal units (BTU) of energy.

We know logically that this can't continue. When it comes to plastic cups the world uses 500 billion disposable cups every single year. We drink water from more than 450 billion plastic bottles worldwide. We use one trillion plastic bags, half a billion plastic straws, and 141 million metric tonnes of plastic packaging.

The world saw a rise in the use of disposable products

during the coronavirus pandemic. Disposable PPE, mainly surgical aprons, gloves and masks were in such demand that frontline workers reported shortages across the world.

Discarded PPE has already been found on the seabed and washed up on shorelines, creating more ocean pollution which will be around for many years to come. Opération Mer Propre, a clean-up group based in France, have stated that there could soon be 'more masks than jellyfish in the waters of the Mediterranean'.

In the wake of coronavirus many companies stopped accepting customers' reusable containers. City to Sea introduced a #contactlesscoffee campaign to urge companies and individuals to continue using their reusables, and showed how to do so safely, in an effort to #choosetoreuse.

In June 2020, more than 100 scientists signed a statement by Greenpeace agreeing that it is perfectly safe to reuse cups, bottles and jars for food, drinks and other groceries, as long as they are thoroughly washed. The statement assures retailers and consumers that reusables are safe, pushing back on claims made by the disposable plastic industry.

What can you do?

It is not a sign of a king or queen to use a towel for a moment and throw it away. A king or queen would use a thick cotton towel, well made and able to be reused over and over. It would be made to last.

It is not a sign of a wealthy person to have a plastic cup. It is a sign of a cheap, throwaway world. A good cup is considered to be a cup that is reusable and has merits over its being merely a container. People express their creativity, their likes and their personality by the cup they choose.

The fashions and personal items of the future will be tailored for a lifestyle that chooses unique and beautiful things of value that last and can be reused.

Look around you, what disposable items are you using every day? Look online for an alternative to your disposable product. Look for products which are labelled 'sustainable' and 'reusable'. At first, reusable might seem more expensive, but over time, as you are not buying the same thing over and over, you will be saving money.

Reject disposability; say no to a disposable world and one-time products like plastic cups.

Take a reusable mug or bottle wherever you go, especially when travelling. There are many reusable cups and mugs available today, including collapsible options that are easy to pop in your bag. Other on-the-go reusables you could take with you are food boxes, cutlery and a napkin.

Encourage your school, college or workplace to be disposable-product free, and make a conscious choice to avoid any throwaway products. You can save money and avoid waste.

Support organisations that work on waste issues: WRAP UK, FareShare, Keep Britain Tidy.

SEE ALSO: Climate Emergency, p. 64; Consumption and Consumerism, p. 77; Ethical Living, p. 135; Litter, Waste and Landfill, p. 218; Your Carbon Footprint, p. 322

DOLPHINS, WHALES AND SEA CREATURES

There are about thirty species of dolphins and whales that live around the shorelines of Britain and about ninety species overall. The smallest is the New Zealand Dolphin, which is only 1.3m long, and the biggest is the Blue Whale, which can grow to more than 30m, making it the largest creature ever to have lived.

Dolphins, in particular, are known to be very intelligent; their brains and sensory systems have evolved and developed to survive in the world's oceans. They've been shown to apply their knowledge and wisdom to survive, to think in an abstract way, to mimic and to learn fast. Dolphins are also social animals; they live in groups and only have a few children or offspring, which they take care of in a group.

When our seas aren't clean these special animals are under threat. For example, a sperm whale relies on food that it finds near the ocean floor, like squid. When the environment is polluted or toxic these incredible creatures, which are at the top of the oceanic food chain, are also harmed.

Dolphins, whales and other sea creatures have the right to live free in the ocean, rather than being fished to near extinction. And why should they suffer for our entertainment, stolen for display in aquariums or dolphin shows? Commercial whaling is supposed to be illegal internationally having been banned in 1986 and yet Norway, Iceland and Japan are continuing to kill 1,500 whales a year.

Our attitude to the ocean and our acceptance of commercial fishing, gas and oil drilling, and international freight shipping is a danger to the lives and sustainability of these beautiful creatures.

What can you do?

Boycott dolphin shows, aquariums and any organisation that thinks that incarcerating animals who once had a whole ocean as their home is a good idea.

Eat only dolphin-friendly tuna. Better yet, don't eat tuna at all. Dolphins swim with tuna and when the tuna is caught some unscrupulous and lazy fishing companies scoop up the dolphins and tuna together. International outrage and campaigning has meant that some major tuna companies have started to label their cans 'dolphin safe' or 'dolphin friendly'.

Support dolphin conservation charities such as the Whale and Dolphin Conservation Society – their wish is that every whale and dolphin should be free to live in the ocean without fear of captivity or death.

Adopt a humpback whale; charities organise gift cards for sponsorship of a named whale. They can even organise visits to the area the whale is most likely to be spotted.

SEE ALSO: Climate Emergency, p. 64; Habitats and Biodiversity, p. 180; Seals and River Life, p. 276; Seas, Oceans and Rivers, p. 278; Water, p. 317; Zoos and Captive Animals, p. 328

DRY-CLEANING

Dry-cleaning uses chemical solvents instead of water to clean clothes. The most common chemical used is perchloroethylene, also known as PERC, a colourless degreasing liquid. Effects from short-term exposure to PERC include nausea, dizziness and even unconsciousness. Long-term exposure to PERC leads to the accumulation of the compounds in the body causing damage to the kidney, liver and immune system, and further studies have found people exposed by working in dry-cleaners to be at an increased risk of developing several types of cancer.

PERC is a volatile solvent and has been shown to contaminate soil, air and water. Its improper use and storage has caused widespread contamination.

What can you do?

Dry-cleaning fundamentally weakens the structure and fabric strength of an item and at an environmental cost.

Most 'dry-clean only' clothes can be simply washed down with a sponge.

Use dry-cleaning only for important items that you can't hand-wash or sponge down. After collecting your dry-cleaning, store your items outdoors or in an unoccupied room so the chemicals and fumes can disperse.

Take a reusable bag with you to collect your items, and refuse their single-use plastic bags and hangers.

If you need to dry-clean your clothes, look for greener, PERC-free dry-cleaners. A quick online search will hopefully find something nearby. The USA-based Toxics Use Reduction Institute is working with small businesses to promote 'professional wet cleaning' using water and biodegradable detergents.

Look for a pure soap for 'delicates' and try hand-washing your 'dry-clean only' items. Natural fabrics like cotton and silk are easily washed at home. Read up about hand-washing delicate fabrics such as silk and wool in order to avoid damaging the fabric or shrinking the garment.

SEE ALSO: Chemicals, p. 54; Chlorine, p. 58; Fashion and Clothing, p. 139; Sustainable Textiles, p. 288

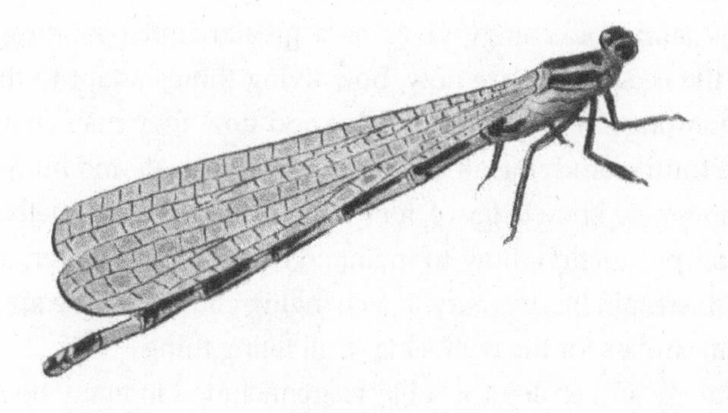

E

ECOLOGY

The word 'ecology' comes from the Greek words *oikos*, home or habitat, and *logos*, study. It means looking at and studying all the relationships and interactions between a living thing and its environment. This includes the study of all living creatures, animals and humans, in their ecosystems.

In this sense, ecology is different from the study of the environment, which is more related to the state of the planet and the difficulties and disasters it is facing.

The study of ecology gives us a greater understanding of how the ecosystems are now, how living things adapt to their surroundings in order to stay alive and how they may change in the future. Studying all living creatures, animals and humans provides new knowledge of, for example, the changing methods of food production, how to maintain clean air and water, and how to sustain biodiversity in a changing climate. These are all crucial studies for the wellbeing of all living things.

Ecology and ecological subjects are included in many books, films, workshops and academic courses, including geography, social studies, politics, mathematics, biology and natural history.

What can you do?

If you are interested in the subject, see if your school, college or adult education centre has a relevant course.

Find your local ecology centre, nature centre or woodland group and immerse yourself in nature. Visit your nearest natural history museum. Or simply walk in the woods. Contact the British Ecological Society for more information.

SEE ALSO: Gaia, p. 165; Habitats and Biodiversity, p. 180; Soil, p. 283

ECONOMIC AND ENVIRONMENTAL BUSINESS

The current economic system is based on a fundamental flaw: the lack of respect for and fair recompense to nature in business.

The wealth of a nation lies in the quality of its soil, in the trees and their fruits and nuts. The wealth of a nation is in its clean, clear rivers and seas, teeming with rich diverse wildlife. The wealth of a nation is found in the numbers of birds and species in the sky, calling to us and screaming their names over and over. The wealth of a nation is measured not in the value of its stock exchange, but in how robust its systems are to create clean drinking water and to provide critical infrastructure and healthcare. It is measured by how many of its citizens have fresh clean water in their homes, enough food and warmth and books to read. That is the mark of a wealthy country indeed.

The wealth of a nation lies in its raw and precious resources, under our very feet. The metals and jewels. The clay or marble. The sand. The peat. The earth and soil. Not in the trees cut down, but in the trees remaining and thriving.

The wealth of a nation can be measured in the way its bees

work, tirelessly sexing every plant and bud so our fruits will grow and soak up the sun all year long. The wealth of our nation can be measured in the number and varieties of apple trees still accounted for that bring forth such an economically formidable package of energy.

The wealth of a nation can be measured in the sheer happiness of its people, in their emotional intelligence, and whether they are well fed with real food, educated with love and treated with respect at work and home.

The wealth of a nation can be measured by the laws of its leaders. The level of respect can be seen in how a leader respects in law the water, the air, the soil, the earth and all sentient beings who cannot speak. How a nation's leadership respects the resources of its country reveals its true nature. The natural diversity of a nation is also its wealth.

When a company lacks respect for nature, it leaves her bare. A company with no respect for nature walks away and remediates only the minimum. When the raw resource is gone the company moves on.

If the imperative of a company is only to make money, then it forgets to respect and put right the source of its wealth. Raw materials for all buildings, for all industries, for all products, come from the earth.

Pharmaceuticals and household lotions are all based originally on the medicines and fragrances of the flower and plant kingdom. Where would we be on earth without flowers!

Where would we be indeed? The natural wealth of our country seems hidden, in native trees, snails and snakes. Our natural wealth seems sometimes ugly, misunderstood, difficult to quantify and harder still to account for.

Yet, account for it we must. If business wishes to maintain its profits then it must adapt and add in at last the true cost to the planet of its produce in all respects. It cannot afford not to.

A corporation can turn its ship around far faster than a law can take effect. It can decide to become an Earth Accounter – a corporation that accounts honestly for its use of the earth's resources.

This kind of company does exist already. There are plenty that have excellent sustainability records, and have appointed managers of environmental standards and others of waste reduction. There are plenty of companies that are doing their bit, taking responsibility for what they do and where it goes at the end. There are companies that have installed solar panels, and reduced its electricity use with a number of measures. There are companies that have good paper and waste policies. There are companies that are zero carbon, that never knowingly kill animals or that are zero waste.

Unfortunately, however, far too many companies feel that such ecological action and mitigation is an imposition on its freedom to make money. This drive for profit shapes everything a company does unless it deliberately allows its motive and law to be rewritten and to account for the earth and all her wealth, thus moving from a financially-driven motive to a sustainable and ethical one.

What can you do?

Support those companies that have taken the wise and important steps towards a sustainable business plan.

Buy Fairtrade, supporting companies that pay a fair wage to workers and have good environmental policies. If you eat foods from abroad – especially cash crops like coffee, tea, sugar and cocoa – buy from companies whose products have the Fairtrade logo.

Cocoa, which is the raw ingredient used to make chocolate, is a cash crop. As the global demand for chocolate rises, producers are struggling to keep up with this demand. In Africa where 70 per cent of the world's cocoa beans are grown, climate change is raising temperatures and dangerously exposing cocoa trees to drought. The cocoa industry is also contributing to deforestation and human rights exploitation such as child labour. If you want to sustain your love of chocolate, it's important to choose Fairtrade organic chocolate from companies such as Green & Black's and Montezuma's. There is also vegan chocolate available from retailers such as Vego.

Reduce your intake of foods from around the world, buying locally and fairly, and supporting nearby businesses first.

Support those organisations that work towards fair trade, empowerment for women at work, trade union rights and sustainable production processes, such as OXFAM #BehindtheBrands, the Fairtrade Foundation, CAFOD and Banana Link.

Buy organic. It's not just better for the environment. You will be supporting businesses that are thinking about their processes and cutting down on chemicals but at the same time making the living and working conditions for those involved more pleasant.

SEE ALSO: Climate Emergency, p. 64; Organic Everything, p. 238; The Development of Humanity, p. 102; Ethical Living, p. 135

ENERGY EFFICIENCY AND ENERGY CONSERVATION

At its most simple, 'energy efficiency' means using less energy to do the same thing; doing more with less and getting as much useful energy as you can without waste. This should be every household's goal: to have heat and light provided at a fraction of the cost of business as usual.

Being diligent and aware of the energy you are using can save huge amounts. It is important for the environment too, as wasted energy causes pollution and exacerbates the climate emergency for no good reason.

It makes sense to heat your home to the same temperature while using less energy. In fact, energy efficiency covers a wide variety of methods to save energy: for example, adding insulation in your roof or inside the walls, using passive solar design, thick curtains, or buying energy-efficient light bulbs.

Compact fluorescent light bulbs use about one fifth as much electricity as an old-style incandescent light bulb but provide the same lux or light. Energy-efficient light bulbs also last longer – sometimes twenty years longer than an incandescent light bulb – so it makes both economic and environmental sense to use them. LED lights are also energy efficient but are not suitable for use in all circumstances.

Passive solar design means a building can use the energy of the sun to heat and cool rooms by absorbing or reflecting the sun throughout the day. It is the least expensive way to heat or cool a home and is used extensively in European countries.

There are straightforward measures that will reduce the overall energy use of heating and lighting in the home. Turning off switches and plugs when an appliance is not in use saves power and money. Sealing doors, air leaks and draughty windows can make a big difference in conserving heat and therefore energy; simple weather stripping or caulking of joints and cracks will save money immediately. It could reduce draughts by 40 per cent.

White goods such as fridges and electrical items have by law to be rated for energy efficiency; a high rating (A+++ is most efficient) means it will use the least energy to do its job and the running costs will be lower and the machine will run at optimum capacity. Some washing machines with smart technology weigh clothes before washing, or reduce water or the temperature needed according to the weather or other factors. These new technological and efficiency measures could make a difference of more than £100 per year per item in running costs and the corresponding energy use. One fridge is available which costs only £14 per year to run.

Energy efficiency is a first easy step to reducing a household's carbon footprint and helping the climate emergency.

What can you do?

Energy efficiency measures should be part of every household and business model. It makes sense to pay less and reduce unnecessary pollution.

Easy things that you can do at home include making sure you have thick curtains, fewer draughts, and roof and walls that are insulated.

The energy-efficiency rating on boilers shows you how efficient or inefficient the machine that heats your home and warms your water is. An inefficient boiler costs perhaps hundreds of pounds more to run than a state-of-the-art efficient one. When you're investing in a new boiler make sure it is the most efficient available.

By watching your energy consumption you can save yourself hundreds of pounds and at the same time reduce your own carbon footprint. Turn off lights when you're not using them, use a water-saving shower head, only boil enough water in the kettle for each cup of tea that you need rather than filling it up completely.

In general: turn off standby items like computers, televisions and games. Install thermostats on radiators so that they can be independently turned down and turned off. Look carefully at the energy rating of anything you have to buy and go for the highest efficiency possible, Grade A+++.

Fit solar photovoltaic panels to your roof: they use the sun to generate the electricity you need, bypassing the use of fossil fuels.

There are plenty of energy information centres and programmes supported by local councils and national government. Some offer free energy-saving light bulbs. Some will help you to do a survey of your house to pinpoint useful places where savings can be made.

Find an energy supplier that uses renewable energy, helps you with energy efficiency at home and has an excellent record on efficiency. Contact: The Energy Saving Trust, the Government's Department of Business, Energy and Industrial Strategy, The Green Age and EnergySage.

SEE ALSO: Climate Emergency, p. 64; The Development of Humanity, p. 102; Fossil Fuels, p. 160; Nuclear Power and Nuclear Weapons, p. 229

E NUMBERS

E-numbered additives are found in sweets and many processed foods, and some are known to cause hyperactivity and in some cases behavioural problems in children.

E numbers are an official form of labelling for food additives. The E prefix before the number shows that the additive has been passed for use within the EU and European Free Trade Association by the European Food Safety Authority.

E numbers fall into seven basic groups. A prefix E1 followed by two numbers means the additive is a colour. E2 stands for preservatives. E3 additives are mostly antioxidants (used to make sure that fat in foods does not go off and taste rancid), synergists (which accentuate the anti-oxidising capability of other substances in the foods), or colour preservatives (which make sure food colour does not fade). E4 are stabilisers, which help things to set and become thick; the E4 list also includes some sweetening agents. E5 are acids and alkalis, used to affect flavours and to prevent the foodstuff from caking – going into lumps.

A number beginning with E6 means the additive is a flavour enhancer. There are no numbers beginning with E7 or E8, but numbers beginning with E9 are glazes and polishers, and from E920 up are 'improving' agents. Additive E925 is chlorine, used in small quantities for bleaching flour, which has been linked to cancer.

Some additives do appear to be dangerous and although authorised for use in the UK by the Food Standards Agency (FSA), they do come with a warning which must be displayed on the food packaging. E102 (tartrazine), 104 (quinoline yellow), 110 (sunset yellow FCF), 122 (azorubine), 124 (ponceau 4r) and 129 (allura red ac) are additives used for colourings which bear this warning. E102 (tartrazine), for example, an orange-yellow colour, is one of several additives that have been connected to hyperactivity in children. Tartrazine is commonly used in the UK although its use is discouraged, yet it has been banned in Norway and Austria.

Additionally, E123, amaranth, also a colouring additive, has been shown to cause cancerous tumours in rats and is now banned in the USA. Additive E621 is monosodium glutamate, commonly recognised as MSG. A flavour enhancer and meat tenderiser, although authorised for use by the FSA, it has been blamed by scientists and consumers for causing hot flushes, dizziness and headaches.

Not all additives are dangerous. E300 is vitamin C, and E322 is lecithin, which works to prevent fat from separating and is actually nutritious in itself.

The difficulty, of course, is knowing which additives to avoid. Maurice Hanssen, the author of *E for Additives*, gives a list of fifty-seven that he suggests should not be eaten.

Ingesting E numbers in food builds up the amount of chemicals in the body. We not only introduce tiny particles of chemical particulates from the air and from our water into our bodies, but add toxins from our everyday food. No wonder people are sick and poisoned. Chemical overload stresses the immune system and it makes sense to reduce the impact.

We know these chemicals are harmful, and yet we continue to eat them. They are made to taste delicious. They are made to lure us in.

What can you do?

We can cut down on the quantity of additive-containing foods we eat. That basically means steering clear of packets and tins of ready-made foods and instead buying fresh vegetables and fruit. Organic produce in season is the greenest and healthiest option.

Avoid strong and strange colours in sweets, drinks and other mainly sweet foods. If it looks unnatural, it probably will be. Learn the E-numbers you wish to avoid and read product ingredients on packages.

SEE ALSO: Additives in Food, p. 20; Allergies, p. 33; Chemicals, p. 54; The Development of Humanity, p. 102; Hormones, p. 196

ENVIRONMENTAL HEROES AND INSPIRATIONAL PEOPLE

As the struggle to understand and implement the changes needed for a new kind of living and a new way to work with our planet rises, so do its heroes. Young women and men from across the planet have been extraordinary, powerful and articulate. They have risen to the challenge and they motivate us in many ways. It is this new generation of environmental activists that inspire us and ask us to do more as we recognise their commitment and planetary actions.

Many people know about Greta Thunberg; she has inspired and challenged millions of people. It started when she began a strike from school one Friday in 2018 and sat outside the Swedish Parliament building. She called it *Skolstrejk for Klimatet*: School strike for the climate, now known as Fridays For Future. Greta was fifteen when she walked out of school saying that the global environmental crisis worried her so much she'd decided to shout about it. She read the science. She didn't understand why politicians weren't doing more and it worried her, like it does many others. Across the world, young people joined her and demonstrated each Friday, asking for more action to protect the planet.

Greta decided to travel across the world, but she opted to reduce her personal footprint by not travelling on an aeroplane, replacing a one-day plane ride with a fifteen-day boat journey. She became a vegan and decided to stop shopping for new clothes; instead, she found her clothes at charity shops. The more she read, the more she changed her lifestyle. She has attended Fridays For Future marches across the world; she has challenged world leaders, industry and religious leaders to read the science, to understand what is happening. 'I'd like to tell my grandchildren that we did it for them, that we did it for their future,' she says.

Ten-year-old Larisa from Russia started picking up litter on the beach with her dog Kovrov. She started a recycling point for batteries at her school and is an advocate for kindness not cruelty to animals. She protects animals whenever she can. She has become known as 'the Russian Greta' because of her enthusiasm and actions and because she looks a little bit like Greta Thunberg too.

Climate activist Arshak Makichyan started a lone strike in Pushkin Square, Moscow, for Fridays For Future. The twenty-five-year-old violinist was jailed for holding a protest without permission. He spoke at the United Nations meeting on the climate (COP 25), on the same panel as Greta Thunberg. He told the meeting: 'I am from Russia, where everyone can be arrested for anything. But I am not afraid to be arrested. I am afraid not to do enough. And I think, and I believe that we can change everything, because behind us there are millions of people, behind us there is the science, and activism is the solution.'

Leah Namugerwa is a fifteen-year-old Ugandan ecological activist. She co-runs the African group Fridays For Future in Kampala. She can be seen with banners at the Kampala marketplace and along the shore of the most beautiful environment she has, Lake Victoria. As well as attending conferences and workshops, Namugerwa has met politicians and ambassadors from foreign nations to persuade them to listen to her cause. 'We as kids, we're not too young to make a positive difference,' she says. 'I want to raise a generation that cares about the environment.'

From the Wikwemikong First Nation tribe in Canada, fifteen-year-old Autumn Peltier was just eight years old when she gave her first ever speech about the people's right to have clean and safe water. Her work with indigenous people has been so inspirational that in 2019 she was nominated for the International Children's Peace Award. Peltier says, 'If we don't speak up now, how much worse is it going to be?' #WaterisLife

It is young indigenous women activists in Alaska that have taken up Climate Justice and called for action on an international level. Seventeen-year-old Quannah Chasinghorse spoke to conventions and federations of native people. 'We shouldn't have to tell people in charge that we want to survive. It should be our number one right. We should not have to fight for this,' she says. With fifteen-year-old Nanieezh Peter, the pair advocated for the Alaskan environment, which is in serious peril due to climate changes.

Bobbi Jean Three Legs is a water protector from Standing Rock Sioux Tribe (Hunkpapa Oyate). She ran a 2,000-mile relay with her brother Joseph White Eyes and about forty other young people to deliver a petition to Washington, DC on behalf of her people. They were claiming indigenous rights on their land, including the protection of their sacred burial grounds and clean water from the danger of unregulated industrial energy pipelines. Bobbi Jean says, '*Mni Wiconi*', meaning 'Water is Life'. The campaign against the Dakota Access Pipeline has received worldwide support. A March 2020 win in an important legal case took the tribespeople closer to their goal of protecting the area when the court insisted on a full environmental impact study before installation could continue which might take many years to complete. #NoDAPL

Sixteen-year-old Rebecca Sabnam is originally from Bangladesh. She migrated with her family to America when she was six years old and is now a well-known climate activist. 'The climate crisis is not just an environmental issue, it's an urgent human rights issue,' she said. 'I am from Bangladesh, a country that exemplifies how interconnected the climate emergency is to racial justice and poverty.'

There are other types of heroes for the planet. Morgan Vague was a biology student at Reed College, Oregon, USA, when she discovered a type of breeding-bacteria that can turn plastic waste into inert enzymes by 'eating' it. Her discovery was described as 'very significant results' by Professor Jay Mellies, who supervised Morgan's research. 'It points the way towards a biological means of degrading plastic pollution.'

Mya Le Thai, from the University of California, Irvine, has unveiled another great invention. She discovered while experimenting with rechargeable batteries that she could increase the cycle of an ordinary battery from 300 times to about 200,000 times by coating the nanowire in manganese dioxide and an electrolyte gel. This extension of the life of the rechargeable battery by such a huge amount is so significant, laptop batteries could last for four hundred years. Her team is working on making a cost-effective prototype for widespread use.

Eight-year-old Mexican Xochitl Guadalupe Cruz invented a solar-powered device to easily heat water made entirely from recycled materials. She provides hot water to low-income families and saves trees! A simple solution is markedly improving the lives of her community.

All over the planet people are coming together and using their skills and resources to create a better world in which we are aware of all the effects we have on water, soil, trees and the climate. Young people are currently leading the way and inspiring others.

Jake Woodier is from the UK Youth Climate Coalition. He reminds all young people that 'children who historically don't have a voice in politics are really thrusting their opinions into the public domain. We are seeing thousands of incredibly intelligent and articulate children who are grasping the severity of the climate crisis better than adults in power.'

Across the UK there are thousands of environmental heroes. People like Amy and Ella Meek, sisters who started Kids Against Plastic in a bid to get schools and businesses to bring what they call a 'plastic clever' perspective to buying and using things. They had been inspired by Bye Bye Plastic Bags, a similar movement created by children thousands of miles away in the small island of Bali.

Licypriya Kangujam is an eight-year-old Indian girl who

started a campaign to ask Prime Minister Narendra Modi to pass the Climate Change Act in 2018. In July 2018 she started a movement called 'child movement' or '*Bachpan Andolan*' to ask world leaders to take immediate climate action.

In 2020, the twenty-three-year-old Ugandan climate activist Vanessa Nakate exposed the racism in the international press. The Associated Press deliberately cut her image from a photo of a group of otherwise white environmental activists who met at Davos to speak to world leaders. Vanessa took the opportunity to remind us again that for Africans and people of colour around the world 'this is something that has been going on for a long while and African activists are trying so hard to be heard'.

Environmentalists in Africa have been speaking of their concerns for as long as those in other countries and sometimes longer. They know the strain that the current system is pressing on their raw resources and emerging industries. They receive little or no attention from the media. Like Licypriya Kangujam and Vanessa Nakate, many activists of colour have found it more difficult, more challenging and more complicated to respond to their emergency.

Another young activist from South Africa, Ndoni Mcunu, a Ph.D. student at Witwatersrand University, reminds us that the climate emergency has already reached Africa. 'Almost 20 million people have fled the continent due to these changes and major droughts have caused almost 52 million people to become food insecure.'

Makenna Muigai from Kenya, Ayakha Melithafa and Ndoni Mcunu of South Africa, Vanessa Nakate of Uganda, and Greta Thunberg of Sweden worked together to get the news of Africa's disproportionate lack of justice on climate issues into the mainstream. They highlighted not just the racial injustices they face but also the economic devastation of their regions despite their communities living a low-impact lifestyle. They told us that Africa contributes only 5 per cent to global greenhouse and carbon emissions and makes up 17 per cent of the world's population and yet they are the most affected and impacted by the climate emergency.

An important factor was that many Africans were relearning to adapt and to use old knowledge and indigenous and local ecological systems for water preservation, seed preservation, harvesting and storage solutions for food. They all wanted more acknowledgement and collaboration with researchers, scientists and politicians on climate solutions.

They inspire us all because they choose to do something positive and share it with their communities. They are interested in social justice and environmental activism. They are unafraid and powerful in their truth.

What can you do?

Even if you are not a well-known international hero you can do your bit and be a hero for your home, your school or your workplace. You have to start somewhere. Not everyone can do all the things our heroes have, but you can all start stepping forward and do something.

If you are not happy about something, then do something about it. Act now: campaign, start a petition and start sharing your concerns.

Whatever you do, do it every day. Even if you can give it only one full minute a day. Do it. Repeat. And then repeat again. Build up your knowledge, experience and interest. Go for it!

Many people find it helpful to use an affirmation to help their mental capacity become strong and match their desire for change. 'I love the planet and I can help' or 'I am a powerful motivator for change' or 'my heroes inspire me to do better and I can'.

This is your planet too.

SEE ALSO: Action, Protest and Empowerment, p. 16; Climate Emergency, p. 64; Conscious Living, p. 74; The Development of Humanity, p. 102; Environmental Stewardship, p. 133; Ethical Living, p. 135; Gaia, p. 165

ENVIRONMENTAL LAW

The most inspiring environmental barrister in the UK, Polly Higgins, died in April 2019. She was just fifty years old. She left behind thousands of people who had signed up to her campaign to have an Ecocide Law, both in British and in International Law.

Polly had discovered that the crime of Ecocide – the deliberate destruction of the natural environment – had appeared in the first draft and many subsequent drafts of the governing papers of the International Criminal Court in Rome, right up to their ratification by sixty countries in July 2002, but was suddenly dropped right at the end of its confirmation process.

Polly Higgins defined Ecocide as 'extensive loss, damage to or destruction of ecosystems of a given territory, such that peaceful enjoyment by the inhabitants has been or will be severely diminished'.

Companies and corporations spend millions of pounds every year lobbying to reduce and nullify our environmental laws. They undermine government when presented with stringent laws to stop their pollution. They employ lobbyists to constantly weaken international conventions and legal arguments that ask them to take responsibility.

Brexit began on 31 January 2020 and Britain moved into

another phase in its history. The most important parts of our EU laws in Britain relate to our environment, especially its protection. Previously the European media called Britain the 'dirty man of Europe', such was our distaste for cleaning up our own environmental mess. Successive governments allowed poor industrial behaviour to continue unchecked. Even when there were laws in place, they often did not prosecute. As a result, rivers and estuaries were dirty and raw sewage and used sanitary towels spilled on to the beaches.

The air was dirty, and still is. Britain's use of fossil fuels is causing deaths in tens of thousands and destroying wildlife everywhere.

The EU collectively agreed to start cleaning the air and clearing the beaches. From plastic pollution to environmental standards, the EU harmonised Europe around a central set of ideas and themes, putting the health and protection of people and the environment at its heart.

It is no wonder there is fear and despair as the Brexit arguments continue and weaken further the chance of environmental excellence in Britain. To do business in the way the government has suggested will mean only one certain thing: that our environmental standards will weaken, and instead Britain will become a lone destroyer, bent on making up environmental rules to suit the lowest common denominator with the cheapest, easiest option dominating, despite public pronouncements to the contrary. Laws may well be re-written to allow only the barest minimum of good practices and safeguards for the environment for the sake of 'business' with other countries outside the EU trading area.

Recent power station plans lay bare the reality behind that public face. The proposal to build Europe's biggest gas-powered station in North Yorkshire resulted in legal action by the campaign group ClientEarth. Essentially, Andrea Leadsom

MP, representing the UK government, overruled her own planning authority in 2019. It had given powerful legal reasons not to build, based on our climate commitments in regard to the Climate Act, 2008. How, the planners asked, could we build such an emitter of pollution with such significant adverse effects and still enforce our own environmental law? The government won their legal case in 2020 against the planners and the validity of the Climate Act was denied. According to ClientEarth, if built, this one plant would emit 75 per cent of Britain's current annual carbon dioxide pollution emissions.

This case will be followed by others, of course; volunteers and lawyers will pace halls, trying to undo such wanton bad behaviour and legal quagmires. Ministers of State will opine and waffle and put off action at every opportunity. They will do little without consistent and repeated reminding of their job, their actual role to protect our lives and our environment. Environmental legal challenges show the truth, and expose how governments fight and deny before they embrace positive environmental responses.

Politicians do fight for what they really want. By following the money, and fighting off the legal challenges, we see that our government is clearly in favour of fossil fuels, nuclear power and fracking at the cost of our health and the sustainability of our nation.

Local people, landowners and council representatives at all regional levels voted and polled overwhelmingly against the fracking industry expansion proposed in northern England at every opportunity. The government decided to change the law so they could overrule those local people, regional government and planning regulations and force on the community the dangerous hydraulic chemical injection into the rocks underground. At each site, nature responded with earthquakes within hours of the fracking starting and it had to be stopped.

There are concerns for the future viability of the industry. This has meant that fracking has recently been declared 'dead' by campaigners at Friends of the Earth in 2020.

The attitude of government and industry is selfish. They waste our time with petty and unnecessary disputes when the survival of our species and the species around us is at risk. How dreadful for this very generation and the next one to find our earth in such a perilous state. The government is denying the climate emergency while at the same time funding and producing plans for large-scale polluting industries they know without doubt cause deaths and add exponentially to increasing temperatures.

There are powerful forces across the world wantonly destroying our earth. Political leaders are actively rolling back the decisions of hard-won legal cases that protect special species, or forests, or common land.

But not everyone feels the same. When the Extinction Rebellion protesters were arrested in London during 2019 they were aiming to highlight their three demands on the climate emergency. What they didn't expect was the remarkable public response from a judge who heard one case with four of the XR defendants: 'This is going to be my last Extinction Rebellion trial for a little while. I think they only allow us to do so many of these before our sympathies start to overwhelm us. When I first started, I was fully expecting to see the usual crowd of anarchists and communists, and all the dreadful things the *Daily Mail* say you are. I have to say I have been totally overwhelmed by all the defendants. It is such a pleasure to deal with people so different from those I deal with in my regular life. Thank you for your courtesy, thank you for your integrity, thank you for your honesty. You have to succeed.'

The Stop Ecocide movement have a simple description of

the way laws should work for us, and on behalf of the planet, too. They remind us that serious harm to the earth is preventable under the following circumstances: when government ministers can no longer issue permits to allow it; when insurers and insurance companies are no longer able and allowed to underwrite such activity; when investors no longer back and fund such activity; and importantly, when CEOs and those in management positions can be held criminally responsible for it. Then, the environmental harm will stop.

There are plenty of environmental laws already; whether we decide to keep them all, improve them or destroy them is one of the most important conversations in the prolonged exit of the UK from the European Union. The coming years will be pivotal in the environmental leadership and necessary response from those in power.

What can you do?

The law seems like a specialist area, remote and distant from our lives. But your local council, county council, Member of Parliament and any other elected representatives work for you and your community. They are the people that are creating this legislation.

If you want the laws to change, to include protections and social justice for people and the planet, it is part of your political responsibility to understand the role of the decision-makers who represent you in your street, your community and in your life.

Become aware of your legal rights as a citizen. Look, for example, at *I Know My Rights*, Liberty's free booklet and guide to the Human Rights Act.

Become an Earth Protector: take a look at the Stop Ecocide website.

You could support the Environmental Law Association or ClientEarth, an organisation directly using the law to fight for justice when the planet can't.

SEE ALSO: Action, Protest and Empowerment, p. 16; Climate Emergency, p. 64; The Development of Humanity, p. 102; Ethical Living, p. 135

ENVIRONMENTAL STEWARDSHIP

Everyone who lives on this beautiful living, breathing planet is part of the earth and has a duty towards its care.

Everyone alive is supported by this planet's incredible ability to sustain life for billions of creatures. What a rich, extraordinary and complex chaos this earth is.

And around the planet there are millions of people who see how precious this beautiful earth is. There are millions of people who protect and preserve the wealth of our seeds; there are people who spend their lifetimes building walls with local stone; there are people building hedges and weaving willow walls. There are those who feed the birds in winter, others still who lay paths and clear away brambles so everyone can walk and see the wealth of our countryside.

There are the thousands who pick up litter; like street mothers, they offer such a dedicated honest service, silently cleaning up after the rudest on the streets. Then look at our stately homes and gardens, jewels of our nation, with tens of thousands volunteering and working hard to open the doors of the best gardens to millions more.

All those who plant trees or clear footpaths, who volunteer in the dogs' home, we salute you! All those who clear the beach of plastic, we thank you! All those who stay up all night to count the numbers of badgers, or hedgehogs, thank you! Thank you, too, to all the staff and volunteers who protect our rare and special species, who maintain ponds and protect rare eggs, and to those who count their numbers and report on their decline.

We are grateful, too, to all those who meet in clubs and pubs, in community centres, in school halls to plan and organise, to fundraise and manage.

Every single piece of this precious earth needs a champion, an earth steward. Millions of people have already stepped up, are you one of them? Where on this earth matters to you? Which piece of land and nature do you love the most?

From the smallest groups to the largest international charities and national and local governmental services, we rely on so many people to keep our environment safe and clean, and able to sustain itself into the next generation. We must encourage and inspire even more to do the same.

What can you do?

Become a silent steward for Mother Earth. Do your bit. Say yes to something.

Environmental stewardship covers a wide range of educational topics; do a course, a workshop. Many UK institutes offer environmental subjects and related courses at A level, degree level and beyond.

Consider volunteering in a group or organisation that protects our natural world.

The Conservation Volunteers have wildlife volunteering opportunities in Britain and abroad. Projects Abroad UK has plenty of animal conservation and protection jobs and volunteering opportunities. You might find a UK or International volunteering opportunity with the International Volunteer HQ, Kaya Responsible Travel or Pod Volunteer.

SEE ALSO: Action, Protest and Empowerment, p. 16; Climate Emergency, p. 64; Conscious Living, p. 74; The Development of Humanity, p. 102; Ethical Living, below; Gaia, p. 165

ETHICAL LIVING

When a person lives within their capacity – i.e. One Planet Living, see page 234 – they can be sure their environmental footprint is what the planet can sustain. Ethical living takes conscious environmental action and merges and expands it with an ethical and moral perspective. This encourages us to be aware of wider issues, such as those that relate to people's justice and fairness. It is no use having a really good environmental policy in a company you support if that company ignores issues of concern that affect our wider society: fair pay, women's equal pay and equality of opportunity, facilities and rights for disabled workers, fair contracts and fair supply chain relationships.

Some companies produce annual ethical reports with their social and environmental responsibilities laid out, which enables you to assess their broader policies.

Your personal ethics will make you the person you are. If you live an ethical lifestyle you will be travelling a path of fairness, awareness, conscious living and doing no harm.

Many vegans and vegetarians feel it is ethical as well as healthful not to eat meat or be part of harming or killing another animal. Many activists practise a non-violent way of life.

The Buddhist faith is a good example of a large religion that has ethical practices at the core of what it is. In this respect there is a path that includes: non-stealing, respect for others, right speaking, right thinking, right actions, mindfulness and effort in one's work and concentration. All these are pure activities that help an aspiring Buddhist to reach a conscious awakening of who they are.

What can you do?

Assess your own ethics. Where do you stand with regard to stealing? Lying? Bullying? Are you a violent or a non-violent person in your core nature? Would you cheat in a test? Would you take a bribe?

People with ethics of truth often challenge those around them. Deciding to tell the truth is a liberating and powerful way to live. Many people are living with constant lies. Choose to speak the truth for a day, a week, a month, and see what happens.

Living your truth is also ethical living. When a person says something, they mean it, they follow it through; they make it a truth and make themselves true to their word.

SEE ALSO: Conscious Living, p. 74; Gaia, p. 165; Keeping Well and Safe, p. 213; Questions?, p. 261

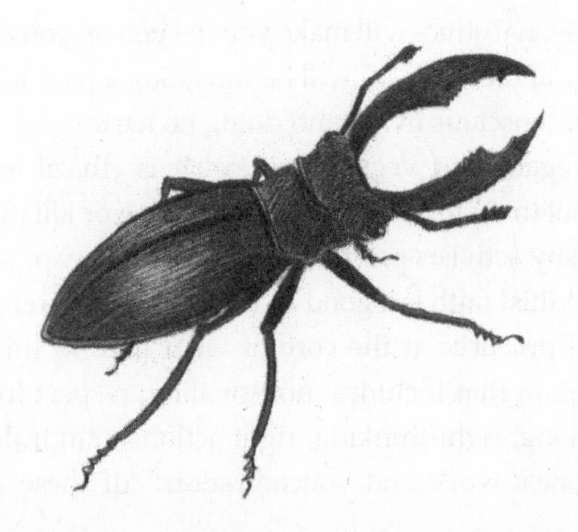

F

FACTORY AND INTENSIVE FARMING

When large numbers of animals are raised and bred in very cramped conditions the practice is known as factory farming. Chickens, pigs and cows are most likely to be factory farmed unless they are reared humanely and with organic feed.

Intensive farming means farmers breed their animals in a small space, which reduces costs and maximises the productive value of the animal. Many environmentalists say that the concentration of animals in a small space not only contributes to the stress and unfair conditions for the animals but also contributes to the climate emergency.

This kind of intensive agriculture makes our meat cheap and our animals suffer. Unless you're eating organically reared or 'compassionate' meat, your meal is likely to have been raised in this intensive way.

The animals are often given antibiotics in their feed as well as other medicines whether they need them or not. The use of antibiotics in farming has reduced recently due to government intervention but industry says it needs to control bacterial infections. Overuse of antibiotics in feedstock can lead to antibiotic resistance. Seventy-three per cent of all antibiotics consumed worldwide are used on farm animals.

In the US there are more than 250,000 facilities that practise intensive farming methods. The desire and demand for cheap meat in supermarkets and restaurants has exacerbated the problem.

The same happens in the UK, where the majority of livestock are called 'zero graze'. That means the cow or pig or sheep will never wander freely in a field. Mega farms are now a core part of the business of farming for the majority of producers.

What can you do?

Contact and support Compassion in World Farming, an organisation committed to bringing about better practices and more humane and compassionate ways of working with animals.

If you eat meat, do you know where it comes from? Find out! Support ethical and organic butchers and ask where they source their produce.

Bypass the intensive farming industry. Eat fresh vegetables and organic foods, locally produced and with concern for animal welfare.

Consider joining Animal Aid, an active animal rights organisation that filmed inside British slaughterhouses and found evidence of law breaking and animals suffering in terrible conditions.

SEE ALSO: Agriculture, p. 27; Climate Emergency, p. 64; Conscious Living, p. 74; The Development of Humanity, p. 102; Environmental Stewardship, p. 133; Ethical Living, p. 135; Gaia, p. 165

FASHION AND CLOTHING

Fast fashion needs to slow down.

Some people use clothes as a way to express themselves, their individuality or their identity; others see them simply as an everyday necessity.

Fashion shapes history and history shapes fashion. Fashion trends have been heavily influenced by social movements, religion, politics and wars. Look at Coco Chanel's invention of fashionable trousers for ladies in the 1920s, after the First World War had shown the advantages of trousers over skirts for comfort and easy movement; or Mary Quant's famous mini-skirt, reflecting the rise of youth culture in the 1960s, and 1980s female power-dressing when women were making demands in the boardroom.

In 2018 the UK fashion, clothing and accessories market was valued at £32 billion, reported by The British Fashion Council. The global industry is predicted to increase in value to an estimated $1.5 trillion in 2020, according to Statista.

We have so much to wear already. According to Eco-Age, households in Britain have an estimated 30 billion pounds' worth of new, unworn clothing in their wardrobes and drawers. And when we do wear it, fast fashion turnaround means that we have about one year of wear from an item before we landfill it. This runaway trend, of fast disposable clothing, chemical-intensive manufacture and the use of plastics in our fabrics, is causing multiple and serious long-term environmental damage.

The fashion and clothing industry is one of the most polluting industries in the world, second only to the oil industry. The report 'Putting the Brakes on Fast Fashion', conducted by the United Nations Environment Programme (UNEP), found that the fashion and clothing industry accounts for 10 per cent of the global carbon emissions as a result of energy used during production, manufacturing and transportation.

For each kilo of fabric produced, 23kg of greenhouse gases are generated, according to *The Economist*'s report 'The environmental costs of creating clothes', which says that half of our fast-fashion clothing lasts for only one year, making it a disposable product, when just fifteen years ago clothing was a more considered purchase and would have been cleaned, fixed and repaired more often.

Synthetic fibres such as polyester and acrylic are made from plastic. The production of these fibres is more energy intensive than that of natural fibres and now 64 per cent of our clothing contains plastic, as stated by Friends of the Earth. They shed microplastics, also called microfibres – millions of tiny particles of plastic waste – during washing and wear. Countries such as China, Bangladesh and India are manufacturing the majority of clothes worn in the UK today. These countries are powered by coal, which is the dirtiest type of energy in terms of carbon emissions and it is directly polluting their atmosphere, causing serious long-term health problems and even excess deaths for local people. Your desire for very cheap clothing is literally killing people across the planet. See 'Air Pollution' on page 29 for more details.

Clothing contains many toxic chemicals; they are used in growing, treating, cleaning, weaving and finishing. Some 23 per cent of all chemicals produced worldwide are used in the textile industry and some 1,900 chemicals are utilised in production according to the report of the EU's research service, EPRS: 'Environmental impact of the textile and clothes industry'.

The fashion industry is also having a negative impact on the quality of the earth's soil resulting in desertification and drought – from the overgrazing of land used to raise cattle for leather, or goats and sheep for cashmere and wool, or from cotton plantations – which threatens agriculture and ultimately contributes to global warming. According to the United Nations, by the year 2030 the fashion and textile industry is predicted to use 35 per cent more land, amounting to over 115 million hectares.

What's more, each year, approximately seventy million trees are cut down and replaced by mass plantations of trees such as bamboo as well as spruce, pine and beech trees, which are used to make wood-based fabrics such as rayon, modal (from beech) and viscose. This deforestation is threatening our ecosystems, our animals and the rights of indigenous communities.

In many of the countries where garments are manufactured, untreated, toxic wastewater from textile factories is dumped directly into rivers. According to Sustain Your Style's report 'Fashion's Environmental Impact', 20 per cent of industrial water pollution comes from textile treatment and dyeing, with 200,000 tonnes of fabric dye polluting rivers and seas every year; 90 per cent of this figure is affecting the waters of developing countries.

A UK Parliamentary Report, 'Fixing fashion: clothing consumption and sustainability', lists the environmental price tag of fast fashion. Creating and growing fibres for clothing is a major use of water in industry worldwide. According to Sustain Your Style, it takes 20,000 litres of water to grow 1kg of cotton, with the industry using a total of 1.5 trillion litres of water every year.

The environmental impact of the clothing industry doesn't stop at the factory gates or at the shop door. According to the charity Sustain, every time we wash a garment made from manufactured fibres, around 1,900 individual microfibres are released into the water, making their way into our oceans. Scientists have discovered that small aquatic organisms ingest those microfibres. These organisms are then eaten by small fish, which are later eaten by bigger fish, and so on, finally introducing plastic to humans in the food chain.

The end of life for a garment is not the end of its pollution. Unless clothing is refashioned, recycled or donated, it goes directly to landfill or is incinerated. Synthetic plastic-based fibres, such as polyester, are non-biodegradable, taking from 20 to 200 years to break down depending on the conditions.

The fashion industry employs approximately 300 million people worldwide, and contributes to the market value within many developing countries, such as in Bangladesh, India, Vietnam, China, Sri Lanka and the Philippines, who rely heavily on fibre production and manufacturing as a source of employment and income.

The industry is often revealed to be paying offshore factory workers (often underage children) less than the minimum wage. They manufacture their garments at factories where human rights, safety and environmental care are at the very bottom of the priority list.

You may have heard of Fashion Week but have you heard of Fashion Revolution Week? Following the 2013 Rana Plaza factory collapse in Bangladesh, when over 1,100 people lost their lives and a further 2,500 were injured, the 'Who Made My Clothes' movement was born. It was set up by Fashion Revolution, a charity with a platform open to the whole fashion industry from brands to manufacturers to individuals. Fashion Revolution fights for environmental protection and human rights worldwide. Orsola de Castro, designer and co-founder of Fashion Revolution, tells us: 'It isn't enough just looking for quality in the products we buy, we must ensure that there is quality in the lives of the people who make them.' Each year during the week of 24 April, Fashion Revolution Week remembers the 2013 disaster.

What can you do?

Own what you wear, how you choose to look, and your personal style. Make it yours.

Reusing clothes, refashioning them and claiming them as yours is a statement of uniqueness and tough ecological standards. The re-fashioning of clothes is the next big fashion trend. Refashioning will make for unique catwalks and dramatic creativity.

Find a pile of clothes you know you will never wear again; they might be damaged, too small, or old but loved none the less. Lay them out on the floor and see which would fit well together. Look for similar thickness of fabric or similar colour schemes and then make yourself a new jumper, or coat, or skirt from a mash-up of fabrics and clothing, using the best bits of what you have. You can use a sewing machine or sew by hand. Get your scissors out and have some fun.

If you really want to buy new, look hard at the new garment you want. Do you know what fabrics it is made from and where it has been made? The 'Made in' label can tell you a lot about an item and the story behind it as well as the fabric it is made from. Is it synthetic, natural or organic?

Do you think the price of the garment is a fair price to pay? Think of the life of that item from the planning, to the manufacturing, to it finally reaching the shop floor. Do you know that everyone involved in the process has been fairly paid?

What about the company behind the brand? Do they have a good work ethic? Do they pay workers a fair living wage? Do they have their sustainability report clearly on their website?

Look for natural, fair-trade, organic fibres or recycled fabrics. These will be less polluting and generally have low water consumption in production.

Don't become a hoarder of clothes – find friends who are the same size as you and bring all your unwanted items together for a swap or sale. Recycling clothing and accessories among your friends can make a great night out.

When you have finished with your clothing and you cannot upcycle or swap them any more, then recycle in charity shops and put the rest in your local council's clothing recycling bins. Shoes can also be recycled in this way. Clothing should not be going into the general waste stream.

Recycle your unwanted textiles, such as unusable items and rag textiles. These can be given to charity shops and some high-street retailers which are then sold on to recycling companies who will recycle textiles into furniture stuffing for sofas and car seats, for example. Be sure to label up the bag of unusable textiles to make it clear and safe for the next person who handles it.

Make money from your unwanted clothes. There are many online selling platforms to do this, along with car-boot, second-hand and clothes-swapping events.

Don't support fast fashion brands. Call them out on their sustainability practices. If you follow any unethical brands on social media, newsletters or home mailings, unsubscribe.

SEE ALSO: Agriculture, p. 27; Animals in the Textile Industry, p. 40; Climate Emergency, p. 64; Conscious Living, p. 74; Consumption and Consumerism, p. 77; The Development of Humanity, p. 102

FAST FOOD, TAKEAWAYS AND RESTAURANTS

There is nothing better than eating out with friends, enjoying the atmosphere and the conviviality of sharing food together. Everyone needs to eat food, everyone likes to have friends and family for dinner. It makes sense that a large part of our social communication is undertaken with a good meal and a drink.

Fast food is readily available in modern Britain; it is delivered, collected and eaten by huge numbers of people on a daily or regular basis. In 2018 the UK fast-food market alone was reportedly worth £15 billion. But it is damaging to our bodies' wellbeing and to the health of the planet.

Burgers, chips, spicy chicken wings and fizzy drinks: there is little quality nutritional value in fast food. Watch the film *Super Size Me* for details. Fast food is generally low in protein and essential vitamins, but high in fat, sugar and salt, with high levels of additives and preservatives. A lifestyle fuelled by fast food is likely to lead to malnourishment, including obesity.

Food production manufacturers offer a wide range of 'meat products': sausages, pies, burgers, meatballs, and many more. However, not everything that goes into these products is the conventional sort of meat you would find at a butcher, or even meat at all.

A '100% pork sausage', for example, must legally be 42 per cent pork, but the remaining 58 per cent can be a mix of many other things, including fillers, additives, fat, wheat and salt.

Mechanically recovered meat is produced by machines that 'tumble-massage' stripped animal carcasses to remove the last shreds of tissue. The resulting 'meat' is finely chopped and ground by machine and then bound with emulsifiers and thickeners to stop the whole ugly mass falling apart. Waste and by-products of the animal, including internal organs, intestines and testicles, have also been found in sausages and other meat products. Fat is added to burgers, pies and pasties before any oil is even considered in cooking, adding further to the overall fat intake if the food is fried. While this is commendable in terms of a zero-waste approach, it is problematic in terms of our overall health. Eating offal, organs and excess fats hidden in these foods increases our heart problems. Meat used from organs is particularly high in saturated fat and cholesterol.

Sugar is also found in high amounts in both savoury as well as sweet dishes, pizzas, sauces, drinks.

Salt is added to raw meat mixtures in large quantities, with the average sausage sandwich containing two thirds of an adult's maximum daily recommended salt intake.

It's often more difficult to find the full ingredients listed on takeaway and fast foods.

Fast food is a global cause for concern due to its continuous damaging impact on the environment. Meat, particularly beef, is the most common product offered for food-on-the-go. Beef production requires extensive land for cattle to be farmed: see 'Climate Emergency', page 64. Of all meats, beef production is the most damaging to the planet.

In addition to the food there is the packaging it comes in: single-use fast-food packaging creates eleven billion tonnes of waste in the UK every year. That's eleven billion tonnes of waste that was used for just a few minutes.

The average packaging from fast-food chains will comprise a paper wrapper and a box for the main dish, a container for the chips, a paper cup, a plastic lid and a paper straw for the drink, a handful of condiment plastic sachets and a paper napkin: all wrapped in a paper or plastic bag. None of the paper can be recycled due to food contamination, and it is unlikely that the plastic will be cleaned and sorted for recycling so it will end up in landfill. If the plastic is polystyrene, as some boxes (usually kebab boxes) are, it is not recyclable at all.

What can you do?

Try cutting down on the amount of fast food you eat, and when eating fast food opt for organic options to avoid additives and hormones.

Practise making quick meals at home that could replace fast

food. It might not be quite as convenient but home-cooked food gives you better nutrition and costs you and the planet far less.

When you cook dinner at home, make an extra portion or two ready for lunch over the next few days. You can also freeze cooked foods.

Write to the major fast-food chains when you see over-packaging and lax environmental standards.

Avoid fast food where you can and always ask for the best environmental product if you do have to eat out. When ordering a takeaway meal from your favourite local restaurant ask if you can bring your own containers from home for them to use for your order. You'll find many independent, local restaurants will have no issues with doing this as you will also be saving them money on food packaging, much of which is plastic.

SEE ALSO: Agriculture, p. 27; Chemicals, p. 54; Food Waste, p. 156; Packaging, p. 240

FERTILISERS

Fertilisers are used all over the world to meet demands for the maximum yield when growing crops for foods, fabrics and oils. There are two types of fertilisers: natural fertilisers, such as manure and bird dung, and artificial fertilisers, which are chemicals. Both can be added to soil or land to increase fertility.

There are significant differences between natural and artificial fertilisers in terms of nutrient availability and the long-term effects on soil, plants and the environment.

Artificial fertilisers

There are three main nutrient groups manufactured as fertilisers, based on nitrogen (nitrates), phosphorus (phosphates) and potassium (potash); and nitrogen-based fertilisers make up the largest portion of the total produced. The British Survey of Fertiliser Practice report 'Fertiliser use on farms for the 2019 crop year' states that nitrogen use on crops is around the same as in the early 1980s, and that it is the most employed of all fertilisers (at 140–160kg per hectare nutrient). However, the use of nitrogen on grass has reduced significantly over the same period to about 50kg per hectare. The use of both potash and phosphates is at an all-time low at around 20kg per hectare each.

Nitrates (from nitrogen) are found in artificial fertilisers. Nitrates can dramatically increase the amount of crop produced, by as much as 100 per cent. Artificial fertilisers are popular among farmers because with them they can meet the high demands from customers due to the growing world population.

According to the Soil Association, many chemical fertilisers contain more than three times the amount of minerals than your fruit and veg actually need. 'This is problematic, because the excess minerals are then washed away by rain and irrigation and find their way into water sources used by both humans and animals, creating a major pollution risk.'

When nitrates get into waterways they start to fertilise the algae and weeds, which then grow excessively. This darkens the water and destroys the natural oxygen balance of rivers and lakes, causing plants and fish to die.

The nitrates that end up in waterways are carried through into the water and food we consume and therefore the nitrogen makes its way into the human body. Nitrogen pollutants in the body are called nitrosamines, and these are carcinogenic. Drinking water has been found to contain high levels of nitrates, which is particularly risky for babies. Baby formula milk that is mixed with contaminated well water has caused nitrate poisoning. Fertilised soils emit nitrous oxide, one of the greenhouse gases, at an increased rate, directly contributing to the climate emergency. Fifty per cent of the nitrous oxide causing the climate emergency is from the nitrogen-fertiliser production process. China produces and consumes about one third of the fertiliser used in the world according to Greenpeace International and Renmin University of China. Overall the global production has spiralled to more than 120,000 metric tonnes per year.

Natural fertilisers

Natural or organic fertiliser is made from plants or animal waste or powdered minerals. Natural fertilisers can include banana peels, coffee grounds, eggshells, grass clippings and even human urine. Other natural fertilisers include cow or chicken manure, limestone, alfalfa meal and kelp.

Natural fertilisers work by releasing their nutrients as part of the ecosystem decay. Natural fertilisers do not kill or attack the hormone system of the creatures around it. As these fertilisers break down they improve the quality of the soil, increasing its ability to hold water and nutrients necessary for successful crops. Natural fertilisers will make your soil and plants healthy and strong. They are also renewable, biodegradable, sustainable and environmentally friendly. They pose no threat of toxic build-ups of chemicals that can be deadly to plants or food crops.

Natural fertilisers are easy to get hold of, like seaweed or nettle, and can be by-products of things from inside your own house as compost, meaning there is absolutely no waste!

What can you do?

Try to avoid nitrates in your fruit and vegetables and buy from organic companies that produce their food without the use of chemical fertilisers.

Grow your own vegetables. You can use natural fertilisers such as compost, which you can make yourself. Compost slowly releases nutrients into the soil. Treat your soil with care and respect.

SEE ALSO: Agriculture, p. 27; Grow Your Own Food, p. 177; Organic Everything!, p. 238; Pesticides, Herbicides and Agri-Chemicals, p. 244; Soil, p. 283

FESTIVALS, CONCERTS AND OUTDOOR EVENTS

Festivals in Britain are huge. Millions of people are involved: musicians, poets, artists, campaigners, crew and happy party-goers. So many people enjoy the UK counter-culture it is now mainstream. From the BBC to major companies, we see that all kinds of people want to be part of the festival energy and excitement. In the past twenty years we have seen major festivals becoming very aware of their environmental footprint.

It's not surprising. When you take an empty field and set it up for a few thousand people to visit and enjoy themselves, you know that at the end you will have to clean up. It's the ultimate party but what is obvious is that accounting for the waste, the immediate environmental impact and the aftermath has a huge cost.

Local councils and emergency services are the frontline. They have the responsibility to negotiate and manage the waste, traffic, emergencies and all the many day-to-day issues that are required so that a town can continue to function even if there are a spare 100,000 visitors arriving for an event nearby.

Festival-goers can make it easy or difficult for the planet. Many people arrive at a festival with good intentions. Being drunk in an open environment then makes them act without inhibitions and without being able to see much else around them. That is why there is so much litter, peeing, drug-taking, so many leftover tents and other generally bad behaviours.

In one sense it doesn't matter how green and environmentally aware the festival managers are – as soon as a large group of people start littering and peeing, the general care of the land becomes a secondary thought. The biggest and most toxic litter problem at a festival is the cigarette butt (for more see 'Cigarettes', page 61).

The best UK green and environmental festivals are ones that have the right policies and core principles:

- A sustainability plan that isn't niche, but involves all the staff and management on every level
- Zero-waste policy; recyclable food cartons, no one-use plastics, no plastic forks, cups. Reusable beer cups etc.
- Recycling policy: 100 per cent waste sorted and recycled, eco plan for food compost and human waste via compost loos
- Mainly plant-based foods: vegetables and fruit, local foods, plenty of full vegetable options
- Stages and cinemas run by renewable energy sources rather than mains power electricity: using solar, wind and bicycle power
- Train-to-bus-to-site transport policy; benefits for cyclists; reduced car use and car share policies
- Solar showers, composting toilets, taps and drinking water stations for refilling reusable bottles
- Learning emphasis on green positive options and alternatives
- A percentage of profits for charities
- Local charities, environmental groups, social and community organisations involved in many areas of the festival production and event

- Tent-reclaim programmes and zero-waste policies; some events fine those who leave waste
- Festivals that have an element of nature and the respect not to destroy it. Messaging during the festival and reinforcing this culture as part of the event are important ways to remind everyone
- Not stocking and selling colas and drinks from multinationals in single-use plastic
- Paying a fair wage and providing good conditions for short-term festival workers

The best truly green and sustainable UK festivals include:

The Green Gathering, Chepstow, Wales
Green Man Festival, Brecon Beacons, Wales
Hay Festival, Hay-on-Wye, Wales
Latitude, Henham Park, Suffolk
Shambala, Secret Country Estate, Northamptonshire
Sunrise, Kentchurch Estate, Herefordshire
Wilderness Festival, Cornbury Park, Oxfordshire

The future of large-scale festivals full of hundreds of thousands of people camping together for days and weeks is now in doubt with new medical and health and safety issues compounding insurance and logistics problems; many will no doubt end entirely. During the coronavirus pandemic many musical and cultural events switched online and played recordings of 'the best of' and festivals became free events enjoyed by many millions of people. Obviously this allowed people to enjoy music without leaving their homes thus reducing pollution. Is this the future for green festivals?

What can you do?

If you are going to a festival make sure you have a zero personal footprint. Travel on the train-to-bus-to-site programmes.

Bring your own reusable plate and mug, be 100 per cent responsible for your own food utensils.

Don't leave anything; don't litter, especially cigarettes or plastic tents. Definitely take your tent and belongings home. Leave the space cleaner than you found it.

Don't stress the environment by peeing or defecating on the site anywhere other than in the allocated toilets.

Avoid killing insects or animals, anywhere. Don't use floating candles, balloons, fireworks or fires.

SEE ALSO: Cigarettes, p. 61; Disposable Products, p. 105; Glitter and Shiny Metallic Decorations, p. 170; Holidays, Travelling and Relaxing, p. 191

FLOWERS AND PLANTS

Flowers and plants make up the most important part of our ecology. Plants produce the oxygen that we breathe. They also absorb carbon dioxide from the air through their leaves. What an amazing miracle! We get to breathe and they suck up our pollution.

Planting flowers can directly help the environment in other ways apart from the miracle of their beauty, perfume and existence. Flowers attract bees and other insects that pollinate and create difference and ecosystem diversity, which increases natural variety and strengthens local species.

Native plants are the best for the soil where you live. When you introduce non-native species this often causes damage, such as soil erosion, or acidification, and the plant would not necessarily be suitable. Sometimes plants from other countries thrive too well in another place, blocking out native plants and creating unwelcome soil imbalances and pH problems. An example is the way that rhododendrons have taken over some wilderness areas in Wales; they created large areas of acidic soil after being introduced in 1673 from Portugal. They quickly take over from smaller, native plants that can become endangered.

Sometimes plants are used to 'soak up' pollution, chemicals and toxins found in the soil. A recent study has begun research on the abilities of the hemp plant to absorb pollutants. It is very adaptable, and the root systems that live near streams serve as filters, taking toxins from the water.

Planting a garden with vegetables and flowers mixed together can bring benefits. Some varieties thrive together, supporting each other and helping to ward off invading insects. Roses love garlic and also chives, leeks, onions and shallots. Other members of the onion family, Alliums, that are grown for their ornamental flowers, can be used by gardeners to repel cabbage butterflies and so protect any cabbages. Garlic planted around an apple or plum tree will repel boring insects; it is best planted with young trees. Nasturtiums are particularly good at repelling some insects that attack cucumbers and squash. To repel cabbageworms plant pelargonium geraniums.

Other insecticidal flowers that produce their own natural insect repellent include asters, chrysanthemums, cosmos and marigolds.

Flowers and plants are amazing as medicine. The crushed, toasted, soaked or liquefied roots, petals, stalks, buds or leaves are often used for specific medicines. Over thousands of years humans have developed sensitive and profound relationships with flowers. Understanding them has helped to create an array of tinctures, creams and medicines that have soothed and saved lives. Many mainstream medicines are produced by mimicking certain compounds found in flowers and plants.

Flowers and plants also are a medicine by their very being, their existence. They are serene and beautiful. People are happy and uplifted when they get a bunch of flowers, especially home-grown ones, as a birthday gift, or a sign of respect and love. Patients are helped and feel loved when friends bring flowers. People find joy and happiness when they have these feelings.

Roses are the symbolic flower of England. And, like the lotus flower, they are considered to hold special spiritual qualities. They have existed for thousands of years and produce one of the most exotic and expensive and sought-after oils in the world. Rose oil and rose water are used in healing, calming, purifying and settling. Roses are the symbol of the Goddess Venus and love.

From the sweetest rose to the tiniest daisy, flowers are a joyful, colourful part of our lives, our love and happiness on the planet.

Buying plastic-potted forced miniature roses or plants is not an environmental advantage. They are sometimes difficult to repot, lasting only for a few weeks or one season and the plastic waste continues to accumulate in our waste stream.

What can you do?

If you have a garden, a patio or even a window box you can start to grow something.

Pesticide-free planting is the way to enhance your environment. Avoid adding to the global problem posed by toxic chemicals. These sprays and pellets are there to kill animals and insects. Find a better way of protecting your plants.

Don't cut your lawn so much either; leave it to bud into life during the spring months, allowing grasses and flowers to seed and spread for the following year.

Use rainwater to feed your plants. Make a pond if you can; it would make the single biggest contribution to the biodiversity and amount of wildlife in your garden.

Plant trees, bulbs, flowers, bushes, any plants that inspire you, that are native to where you live, any plant that makes you happy.

Local garden centres have lots of information; they can show you which plants are native. The internet is packed with information about every plant that exists. You can find so much history, as well as planting and seeding ideas and tips on how to garden in the greenest way. Ask other gardeners for their tips, their seedlings and cuttings. Many gardeners will be happy to swap plants with you.

Join an organisation that monitors, protects and raises awareness of issues relating to wildlife, wildflowers, plants and native species protection.

Join a local gardening group if you don't have any space yourself. Offer to pick weeds and learn about gardening in a home where there is someone who can't garden, such as an elderly neighbour or relative. There are also 'allotment' programmes in local councils that offer a strip of earth on a community plot for a small fee.

If you have to buy flowers from a shop look for those that are Fairtrade Foundation approved. Most major supermarkets have joined this programme.

SEE ALSO: Agriculture, p. 27; Gaia, p. 165; Grow Your Own Food, p. 177; Soil, p. 283

FOOD WASTE

Have you ever been so hungry that you have piled up your plate with food, then discovered that you couldn't finish so had to throw some away? Or have you ever cooked a little bit too much, popped it in the fridge, forgotten about it and when you discovered it again a week later had to bin it? Globally the Food and Agriculture Organisation of the UN counts 1.3 billion tonnes of food wasted every year.

As a nation, the British spend billions of pounds every year feeding ourselves and our families. You wouldn't throw money away, so why throw away food?

It all starts with the production of food. Did you know it requires 80 litres of water to grow a single orange, 320 litres for an avocado and a staggering 790 litres of water for a bunch of bananas? According to the World Resources Institute, we use 70 per cent of all fresh water on food agriculture and 24 per cent of the food produced simply gets thrown out and ends up in landfill as food waste. That represents 45 trillion gallons of fresh water lost every year.

Over a third of all food produced never even reaches our tables. Some of the produce grown does not even leave the farms if the food is seen as less than perfect, which means popular supermarkets will not buy it. Additionally, any food that has been damaged in transit will be disposed of before it reaches the shop.

The food we throw in general-waste bins is destined for the landfill. When food ends up in landfill, it releases methane gas, which is up to one hundred times more damaging in the environment than carbon dioxide. If the world's global food waste were a country, it would be the third-largest emitter of greenhouse gases after the US and China.

Sharing, recycling or composting our food waste could cut our CO_2 emissions and reduce our carbon footprint by up to 25 per cent.

One method of recycling our food waste is by anaerobic digestion, which is usually facilitated by a food collection service provided by your local council. Anaerobic digestion is the process of putting food waste and industry compostable materials into a tank without oxygen, where the waste is then broken down by microorganisms. As the waste breaks down it produces a biogas, which is then collected and used to generate electricity. By managing our food waste in this way, we can also make animal feed.

In the UK, 1.3 million tonnes of food waste goes through anaerobic digestion, creating enough energy to power 200,000 homes each year, which is worth more than £220 million to the economy according to the consumer organisation Which?

Home composting is an easy way of changing food waste into useful compost. Everyone with a garden should have a compost heap. This is a great way to reduce the environmental impact of your uneaten food and help your garden with high-quality mulch for your own vegetables. Some local councils send the food waste collected at the kerbside for composting rather than anaerobic digestion. Commercial-scale composting can safely include items like meat waste and cooked food that are not suitable for home composting.

A report released by the Department for Environment, Food and Rural Affairs (DEFRA) confirmed that UK households throw away around 7 million tonnes of food every year. The average household throws away 20 per cent of food purchased, equating to approximately £700 a year per household. Of the 7 million tonnes thrown away, 5 million tonnes are edible food scraps, such as the last few bites from your plate that you couldn't quite manage.

That is bad enough, but 250,000 tonnes of the 7 million total is untouched – that is, perfectly edible food that could be used to make up to 650 million meals for those in need. The hunger charity FareShare say that more than 8 million people in the UK are struggling to afford to eat and that this number has risen since the coronavirus pandemic. Some food banks saw a 300 per cent increase in their applications. The UK Food Foundation said that during the lockdown more than 1.5 million people went without food for a whole day because they had no access to food or money.

Global food production must increase by 70 per cent by 2050 in order to meet the demands of the growing world population according to the Food and Agriculture Organisation of the UN. If we were more conscious of food waste we might meet that target without the need for more agribusiness and the effects that whole business has on the earth. The solution, mathematically, is to support small-scale organic farms and community gardens. When people start to grow a higher percentage of their own food close to their dwelling, the production can rise to meet the global demand.

What can you do?

By using up every edible morsel of your food, you're doing your bit to look after the environment; imagine what we can achieve if we all make a change!

We need to be less wasteful with our food. Organisations such as Oddbox are working towards finding a market for the produce the growers otherwise have to discard as unsaleable, delivering fresh fruit and vegetables that are classed as 'wonky' or 'surplus'. And online apps such as Olio, TooGoodToGo, Karma and ReducedFoodLondon are all doing what they can to reduce the amount of edible food sent to landfill and at the same time help feed people at a lower price or for free, donating what is left at the end of the day to charities and food banks.

We too can do our bit to cut the amount of food destined for the landfill, first by buying and cooking only as much as we need – writing a meal plan for the week can help with this – then by freezing or sharing any excess.

Share it. Share your unwanted food; don't let it go to waste. Speak to your neighbours; get to know your community. There are food-sharing apps these days including the Community Fridge Network (hubbub.org.uk) – get on them and start listing!

Understand the difference between 'use-by' and 'best-before' dates: 'use by' is about safety – food isn't safe to eat after this date according to the producer. 'Best before' is about quality – foods are perfectly safe to eat after this date, they just might not taste as good.

Check your fridge is at the correct temperature: between 0 and 5°C. This makes it more energy efficient.

Keep bagged or boxed salad and spinach leaves fresher by putting a piece of kitchen towel in the bag or container – it will stop the leaves going slimy.

Compost inedible food waste and use it in your garden. There is a huge amount of advice online and often councils offer free or cheap compost bins.

SEE ALSO: Agriculture, p. 27; Climate Emergency, p. 64; Grow Your Own Food, p. 177; Litter, Waste and Landfill, p. 218

FOSSIL FUELS

At the 2020 World Economic Forum in Davos, Greta Thunberg and twenty-one young environmental activists asked the forum to make sure that they stop investing their money in fossil fuel companies: 'Today's business as usual is turning into a crime against humanity. We demand that you play your part in putting an end to this madness.'

Burning fossil fuels such as coal and oil is the biggest single reason for climate change. The earth is facing catastrophic conditions as the temperature rises, ice is melting and forests are burning.

In order to make a difference and to reduce the global temperature overall, fossil fuels have to be kept in the ground. Unfortunately, governments around the world are continuing to invest money in and pay subsidies to companies producing fossil fuels. In fact, the International Monetary Fund says that the big investment banks have invested £535 billion in fossil fuel companies since 2015.

In the recent past, the fossil-fuel industry and certain governments, including the British one, have tried to promote and proliferate the use of hydraulic fracking to obtain shale gas and oil from deep in the ground in the rocks below us. Fracking is a controversial and very damaging process, often causing earthquakes and earth tremors. The main concern over the high-pressure blasting of chemicals into the fissures in rocks is its long-term effect on the water table and the stability of the earth itself around the plant. The process is more damaging to the climate than other drilling processes as it needs more energy to find and extract than conventional gas exploration. Greenpeace says that if the same amount of financial support and economic advantages were given to the renewable energy industry, Britain could provide more than half of our energy need with less money and negligible carbon pollution.

During the coronavirus pandemic, global carbon emissions fell by the biggest ever amount and there was a similar drop in demand for fossil fuels. As the world stood still, the share price of oil and energy stocks began to crash as demand hit an historic low. According to data commissioned by the *Guardian*, coronavirus cut billions of barrels of oil, trillions of cubic metres of gas and millions of tonnes of coal from the global energy system in 2020 alone.

The planet's coal, oil and gas deposits have all taken hundreds of millions of years to form – which is why they are called 'fossil' fuels – and exist in a finite quantity. They contain very high amounts of carbon, which is released into the atmosphere as carbon dioxide when they are burned. So on both counts they are an unsustainable and dangerously polluting source of energy. It is clear that continuing to invest to make fossil fuels cheap is a dirty business and unsustainable for the planet and for the economy.

Some estimates suggest that coal, gas and petrol are the most subsidised forms of energy that we have ever had. Some people are making a lot of money from coal and oil. Oil – petroleum – has been destroying the planet since the first large-scale well was struck, in Pennsylvania in August 1859, a relatively recent event in our planet's history. You can directly correlate the increase in CO_2 in the atmosphere with the increase in fossil fuel use. We were warned then, in the 1860s, that there was a cause and effect.

Now we must choose other ways to heat our homes and run our cars.

The fossil fuel industries have fought bitterly to keep their money and influence: they comprise three per cent of the global economy and make over $87.5 trillion a year. Of course they want to keep the money. These companies have directly polluted the planet with oil slicks and contamination of wild areas and many have simply walked away and refused to pay for the damage.

Burning oil, coal and gas to run industry, manufacturing and create buildings, roads, trains and international shipping, pollutes by airborne emissions. Our need for energy systems, including for heating and cooking, means millions of people rely on fossil fuels every day. They have polluted the planet indirectly, too, through every car and aeroplane in the world.

Even electric cars have a negative environmental impact in manufacturing and production, and especially if the energy source for recharging the electricity comes from fossil fuels.

Air pollution from fossil fuels is also killing many people. According to the World Health Organization, 4.6 million people die each year directly from air pollution and 7 million people overall from related illnesses including from indoor air pollution such as from using kerosene. That is more people than die in car accidents globally. One in six global deaths has been linked to air pollution (see 'Air Pollution', page 27).

During one week in April 2020, Sweden became the third European country to phase out coal power as it shut its last active coal-burning power plant two years ahead of schedule, in an effort to achieve an energy system based entirely on renewable sources. Earlier that same week, Austria had announced the closure of its last coal-fired plant, joining Belgium in completely removing coal power.

Fossil fuels are over. It is unsustainable to continue in a business-as-usual manner.

We can't say we haven't been informed. People have known for a long time. We are hypnotised into thinking it will all be okay.

We know how dangerous they are and the effects they cause. Those shouting to allow these industries to continue have no moral, environmental or economic argument left.

What can you do?

Stop using fossil fuels wherever you can. Reduce your need for oil, gas, petrol and coal. Switch your electricity provider. Buy environmentally-friendly energy generated from renewable sources, from companies such as Green Energy.

Consider the ideas offered throughout this book for ways to

reduce your pollution and footprint. Change your lifestyle and be a green activist to be part of the solution.

This is an emergency. Ask what your local politician is doing.

#SixthMassExtinction #actnow #climateemergency #ourhouseisonfire #ClimateFacts

SEE ALSO: Air Pollution, p. 29; Climate Emergency, p. 64; Energy Efficiency and Energy Conservation, p. 116; Transport, p. 294

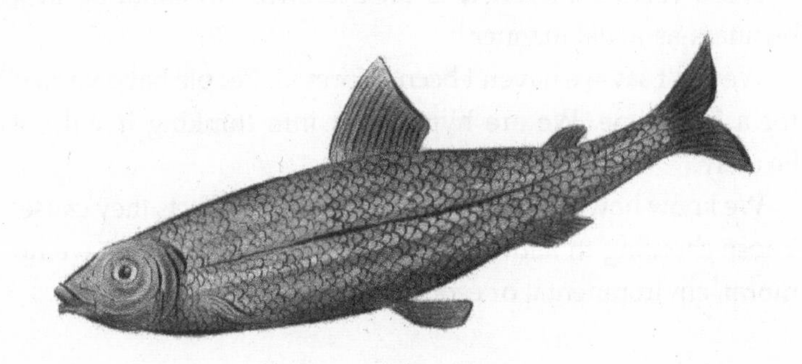

G

GAIA

Gaia is the ancient Greek name for Mother Earth.

Our earth hangs in space, surrounded by a thin layer of atmosphere. From a distance, it looks like an exquisite blue, misty pearl. It is unique. No other planet in the solar system possesses exactly the same blend of gases to create life as we know it.

The biosphere, 'the sphere where life exists', is the thin layer that includes the lower atmosphere, the oceans and the soil, which supports all life. The earth sustains itself by a complex combination of seemingly chaotic systems and cycles. Climate patterns, the cycle of energy, ocean currents, winds and hundreds of other systems all work together, making life possible.

The theory that everything that lives on the planet is a self-regulating, living, complex, evolving system is called the Gaia hypothesis. This hypothesis has been developed by Dr James Lovelock, a British scientist, though it has been part of religious stories and myths for thousands of years. We know that this planet and all that happens on it is interconnected.

One way to think of the planet is as a living body made up of mostly water, just like we are, with different organs in the body represented on earth by the mountains and volcanoes, oceans, soil and greenery. A Scottish scientist in the eighteenth century, James Hutton, described the planet as a 'superorganism' when he compared the movements of the soil through plants and animals in the carbon cycle to the movements of blood flowing and circulating around the human body. We humanise this living organism because we can then identify with it and have sympathy and empathy; then we can connect.

If the planet is like a human body, she can also be sick, and the pollution, waste, radioactivity, chemicals and dangerous gases emitted from industrial processes into the air, soil and water all contribute to the sickness. Understanding Gaia means understanding that the survival of the plants and trees and wildlife that live on this planet with us, is crucial to our own survival. We are all connected.

Human beings have directly caused the problems facing us, so we must help to heal the planet for the future if we want to survive and live peacefully on it.

We have used 'she' to refer to the planet and Gaia because early stories about the beginning of the earth repeatedly use the term 'Mother' for the planet. The first images in religion and art devoted to the Earth Mother appeared some 30,000 years ago, and holy symbols, relics and statues can be found all over the world representing this Earth Mother.

Gaia is also known in every race and culture, though by thousands of other names.

What can you do?

Almost every section in this book will help you live in greater harmony with the planet. You can also read more about Gaia and stories of the Great Earth Goddesses.

What is your personal relationship with the earth? Spend time in silence with her and see what happens. How does your communication grow as you listen and notice nature and its voice?

SEE ALSO: Climate Emergency, p. 64; Ecology, p. 112; Habitats and Biodiversity, p. 180; One Planet Living, p. 234; Species Extinction, p. 285

GIFTS, PRESENTS AND GREETINGS CARDS

Every year millions of people give presents to each other. It is a shorthand way to say 'I Love You' or 'thank you'.

Our great economic wealth means that we now constantly overspend, and so there is a risk that our gifts become less about communicating love or thanks, and more about the latest fashion, the latest gadget or electronic device.

Over the Christmas period each year, an extra 30 per cent more rubbish is produced and thrown away according to local councils who deal with the mountains of extra trash. This additional waste will be in the region of 3 million tonnes. What is more, a UK consumer survey suggested that £42 million of unwanted Christmas presents are thrown out into the landfill each year by ungrateful recipients.

Adding to this waste mountain, there is the waste produced throughout the year from birthdays and other monetised gift-giving events, such as Valentine's Day, Father's Day and Mother's Day, and other religious festivals and celebrations.

Yet gifts don't have to be physical objects purchased brand new from a shop or online store. You shouldn't stop saying I love you or thank you to someone who has been kind or is celebrating. There are wonderful opportunities in life to express gratitude, to share your abundance or to simply show your love.

There is a secret universal law of abundance that suggests that everything you need in life will come to you and the universe in its abundance will supply it. That would make all of life a gift, which it really is.

What can you do?

The most environmentally-friendly gift of all is your time. Spending time with a loved one in the park, going for a walk in a woodland or forest or on the beach makes memories, is fun and is completely free.

Gifts can be experiences such as a fun day or evening out or an activity. You can also give up your time to support someone: maybe your friend needs some help to complete a task and an extra pair of hands would be greatly appreciated.

If you want to give a physical thing, but still want to be as eco-conscious as possible, you could try your hand at making something. Renewing old things or upcycling can be fun. Or get crafty and create something like a drawing, or explore your sweet side with some baking.

When the time is not available to create a gift and you need to buy something, first opt for high-quality used and second-hand purchases; try charity and antique shops, or online selling groups through social media platforms. If those options don't work out, look for ethical, cruelty-free, organic, locally crafted gifts.

Stop buying rubbish gifts. It is much better to have a proper experience with someone. If you want to buy something useful, ask the receiver if there's something they need.

Consider 'adopting' an animal from one of the environmental and animal charities who are fighting to protect them in the wild. Or help a family in need by sending books, or providing clean water, or look for 'feed a family for a year' schemes.

If you're wrapping a gift, always opt for recyclable paper, combined with a biodegradable paper tape. Even better, how about trying no tape at all? Instead, use twine or reused ribbon.

Have you heard of the Japanese wrapping technique, Furoshiki? It involves using fabric to wrap gifts; this fabric can also form part of the gift, of value to the receiver in itself rather than merely decoration. No spare fabric to hand? How about using a pretty towel or scarf from a charity shop, and get creative! Check out Furoshiki tutorial videos online.

Rather than buying greetings cards, you can send e-cards or meet up with or call your loved one to tell them what you would like to say. It takes one tree to make the quantity of 220gsm board needed for just 3,000 average-sized greetings cards. If you do buy cards, look for ones printed on recycled board, or plantable cards which are filled with flower seeds.

Try upcycling any received cards, especially those with lovely images, by making gift tags or new cards. Make a frame for those you really like and they are upgraded into a gift. Gift them onwards or use the images by sticking pictures on card or boxes using decorations and crafts. Wrapping papers can find a new use as decoupage decoration on jars, furniture or household items.

Don't stop sending gifts and cards filled with messages of love and hope to those you love. Resolve now to do it with environmental awareness – that's what will make the difference.

Take time to make things with love in your heart; the outcome will be filled with that energy and be more important, more valued and treasured than any purchased trinkets.

SEE ALSO: Consumption and Consumerism, p. 77; Disposable Products, p. 105

GLITTER AND SHINY METALLIC DECORATIONS

It's your birthday or you're going to a festival, what do you think of? Glitter and shiny metallics, of course!

Remember that time you received a card covered in glitter? You probably found little starry pieces months later.

Glitter is a form of microplastic. Glitter and shiny metallics are usually made from a combination of aluminium and plastic, in most cases polyethylene terephthalate (PET). When PET plastic starts to break down into smaller pieces it releases toxic chemicals that can disrupt human and animal hormones.

You can also get glitter made from a combination of metal and glass; although not as common, it is equally polluting.

Glitter and shiny metallics are impossible to recycle and therefore most of the time they go directly to landfill. Pieces of glitter, however, end up spread all over fields, parks and gardens, meaning tiny particles get into the food chain.

Metallic wrapping paper and metallic pieces within packaging and cards are also not recyclable.

Glitter and metallics are also used in arts and crafts, in party decorations such as confetti and in make-up such as nail polish and eyeshadow. Some people even use pure glitter as stand-out make-up.

Conventional glitter and shiny metallics are non-biodegradable and can take hundreds of years to decompose. In fact, glitter has had such an adverse environmental impact that scientists and campaigners are calling for a global ban of plastic glitter. As of today, no countries have implemented a ban of these decorative plastics. However, we have seen various steps taken to control the use of glitter and a rise in sustainable alternatives.

The Association of Independent Festivals has a

movement – 'Drastic On Plastic' – and in 2019 we saw sixty-one UK music festivals banning attendees from wearing glitter to try to slow down the problems of its being found on farmlands and parks.

Additionally, a number of nurseries and schools have banned the use of glitter within art classes.

But this doesn't have to signal the end of your sparkle. There are a growing number of brands creating biodegradable, eco-friendly glitter, such as EcoStardust and Eco Glitter Fun. This is plant-based; although to have zero impact on the environment it needs to be disposed of in compost.

An alternative to plastic confetti is to use leaves that have fallen from trees and plants. Use a paper hole punch and create lovely homemade confetti that will naturally degrade.

SEE ALSO: Festivals, Concerts and Outdoor Events, p. 150; Gifts, Presents and Greetings Cards, p. 167; Plastics, p. 253

GRAFFITI

When a young person paints their name on a wall or fence they are showing us much more than a special tag name. Graffiti is a complex social mark, mainly carried out by children in distress, in need and suffering emotional harm.

Children can start graffiti at about the age of eight or nine. They are young. They often don't realise the effect of their markings on bus stops and in school. They start with biros or whiteboard pens stolen from school. They work their way up to marking their tag on high walls, bridges and on top of buildings. Many use aerosol cans.

In the UK today, graffiti is directly responsible for many young people's deaths. Children and young people can die on railway lines, they can be electrocuted, or have fatal falls from high walls or buildings. In February 2019 a coroner found that three young men died while hiding from an oncoming train after carrying out graffiti in south London.

Children who are unsure of who they are feel the need to mark their territory. Research has shown that when young boys especially are going through their teenage years they are more likely to experiment with graffiti if they are in trauma or distress and have difficulties in communication or lack the support structures at home or school.

Boys link up together and go out to dangerous places in the dark to mark the walls and surfaces where people might see the graffiti of the gang. But there is an addictive aspect to it; the adrenalin and nature of getting away with an illegal act pull young boys back night after night.

One in twenty graffiti offenders is likely to be a girl.

The environmental damage done by graffiti is practical and costly. Companies in the UK today spend around £1 billion each year to remove graffiti. It costs £10 million to replace glass on land and underground trains that has been scratched and another £2.5 million to clean up aerosol graffiti on station platforms and bridges.

Graffiti also blights an area, making it feel more unsafe. Women especially feel unwelcome and unsafe in an urban area covered by graffiti. It is not a form of art. Graffiti is a cry for help. Instead of criminalising young people for graffiti – it is illegal under the Criminal Damage Act, 1971 – we should be supporting them to find some love and connection before it is too late. We should be helping them to integrate in the community. We should be giving them space to explore themselves in a safe environment, such as in youth clubs and well-resourced playgrounds and after-school clubs.

Banksy is an artist whose work is found on walls and in public places across the UK. The artwork of Banksy is well-regarded; he is known to have attended art school and university. He has produced some of the most famous and articulate work that addresses modern British life, its craziness, its social unease and its humour in dark times.

Banksy is now an internationally known artist and his work is very valuable. Banksy is not a 'graffiti artist' although he is regularly called that and he is idolised as if he is. Rather he is a modern political artist who uses the urban environment to show his work and explore how the urban environment is also our social and political canvas.

What can you do?

Don't start graffiti. Don't condone it either; don't let your friends or family graffiti.

Love where you live.

If your friends are thinking of painting graffiti help them to see how it would make your town worse not better for their tag.

Plant a tree, or a bulb, or some wildflowers without permission; that's guerrilla planting.

Walk in nature. Go swimming. Do something other than destroying the environment around you.

Talk about it. Get some therapy. Find a wiser person to talk things through with. Tell someone you love.

Instead of using spray cans, some would-be graffiti artists are creating masterpieces of their own on walls by removing the dust and dirt to form their design, calling it Reverse Graffiti.

SEE ALSO: Environmental Law, p. 128; and Environmental Stewardship, p. 133

GROUP ACTION AND PROTEST

> 'And yes I know that we need a system change rather than individual change, but you cannot have one without the other. If you look through history all the big changes in society have been started by people at the grassroots level. No system change can come without pressure from large groups of individuals.'
>
> Greta Thunberg, 2019

When a group of people join together to make a difference, they can do extraordinary things. When you get together with friends, relatives, work colleagues, those whom you learn with and play with, you can influence each other. There are thousands of environmental groups that you can join and work with to make changes in your local area. You can meet up with many people to work together or you can start a group yourself.

Be realistic, you won't be able to do everything! Choose a target, a goal, and work together to make it happen.

Article 10 of the Human Rights Act says you have the right to an opinion. You have the right to free speech. In Britain we can speak without fear. We can hold opinions that might differ from others' but we hear their views.

Your comments and your views matter. Every time someone writes or telephones a media outlet such as the BBC or ITV to make a comment on a programme or on some news coverage, for example, five hundred people are deemed to think the same. In effect, your letter, your call, your tweet matters and you are potentially representing another five hundred people who share your view!

It is important to be heard. Your Member of Parliament represents you and makes laws and agreements for the whole country. You can meet her or him, as they are available in 'surgeries' where you live, where you can talk to them about your concerns. They have to meet with you and they are obliged to discuss their environmental policies and what they are doing in parliament about them.

When you meet in a group you choose what to do together. You can choose to protest and to demonstrate. You can choose to take positive, particular action. You can write letters. You can boycott a particular product. You can do a litter pick, a beach clean-up. You can do so many things together for good.

Article 11 of the Human Rights Act says that you have the right to protest peacefully. You have the right to join with others to peacefully protest in marches, demonstrations and pickets. It is your legal right.

Picketing and holding placards with your views is not illegal. When you protest for the environment you are joining with others fighting for democracy and the planet. Liberty, the free speech organisation and pressure group, says, 'Protest is fundamental to democracy and our ability to stand up to power. We corrode the right to protest at our peril!'

You also have the right to join a political party or a trade union and express your views through them too.

Extinction Rebellion (XR) grew in prominence during 2019 with their very unusual form of protest. They decided to allow themselves to be arrested, and in huge numbers in carefully organised, non-violent protests of direct action. The plan was that the group would have a say, a place in court, to voice their concerns in front of a judge and at the same time take up space and time in the court system. They did not expect so many judges to throw out or dismiss their cases, to agree with their concerns and to add their voices to the serious professionals in the UK who want immediate climate action. Out of the thousands of people arrested for protesting in favour of the earth, only a handful have faced serious prosecutorial peril.

XR have three central demands:

1. Government must tell the truth by declaring a climate emergency.
2. Act now, immediately, to reduce carbon emissions by 2025.
3. Organise Citizens' Assemblies to manage and organise the transition to a zero-carbon economy.

There are many other kinds of group action. People can work together when they have a common goal. Sometimes it is in response to an all-out emergency – a forest fire or a river overflowing and flooding. Sometimes it is to support others around them who are poor, vulnerable or sick.

The Good Goddess Project, for example, started in Preston, Lancashire. A few women had heard that in Roman times, Bona Dea (meaning the Good Goddess) was honoured by women who had much time and wealth, by their sharing it or raising funds for women who didn't have resources and who were often in desperate conditions. It was a story of women supporting women.

This inspired the women in Preston to get together and make collections for women in the local shelters, refuges and transition homes. The group started by supplying small bags with personal care items: soap, sanitary towels, flannels and fresh underwear. Others knitted blankets to give as gifts to the women at Christmas.

The project went from strength to strength and now hundreds of women are involved. The women themselves have also benefited from women's circles, group meditations and yoga, singing and cultural sharing with each other.

What can you do?

Animals frequently work together in herds, birds in flocks. Creatures of all shapes and sizes work together in tiny ecosystems. The resilience and group activity of a family of ants is simply stunning. When danger strikes, animals communicate, they work together and help each other.

You can choose to be helpful in your community. You can choose to do your best and support someone else who is struggling and in need. You can choose to join a group and work hard to realise your goals.

Join The Good Goddess Project and help women in your local area by doing something good for vulnerable women who need something good to happen to them. How can you help?

Know your rights as a group, and when you undertake actions and protests. See, for example, Liberty's free booklet, *I Know My Rights*.

The National Council for Voluntary Organisations has registers of local and national charities, training, and blueprints for new groups.

XR have local groups across the country. #RebelforLife #ActNow

SEE ALSO: Action, Protest and Empowerment, p. 16; Climate Emergency, p. 64; Environmental Heroes and Inspirational People, p. 121; Environmental Stewardship, p. 133; Ethical Living, p. 135

GROW YOUR OWN FOOD

Growing food is not too hard. It's a great way to be green and to touch the earth and love the earth. During the coronavirus shutdowns millions more people started to grow their own food in tubs, in their front yards and gardens and on their windowsills and patios.

Start with a small patch in your garden or a window box. You can fill tubs, containers and pots inside. Pips and seeds will grow. Stones will sprout. There is so much you can do to bring greenery of every kind.

Growing your own food is a great way to begin to be self-sufficient, bypassing the supermarket system, the packaging, the pesticides, the advertisers and the big multinational companies that sell you their products. You will have satisfaction and control over your own food and at a tiny cost.

You will also have a valuable community commodity to barter, swap or share, and to offer in abundance. When you grow your own food you have a new kind of wealth.

People often start to grow something simple to save money and once they see the sprouting and the budding of fruits and vegetables in front of their eyes they are hooked! It's fun to do, doesn't take too long and adds a whole new form of self-sufficiency and depth to your life.

Permaculture is a practical ecological philosophy for growing your own food (but also a life philosophy) that is about living lightly on the planet and making sure that the growing and agricultural practices we have will be sustainable for many generations into the future. People who practise permaculture try diligently to make the soil better. They work hard to make sure that water is clean. The idea is also holistic, joining people, ecological systems and social justice together.

To make permaculture food-growing work, composting, mixed and companion planting and soil regeneration are important. Permaculture food-growing doesn't use pesticides or chemicals; the system uses other plants as natural insect repellents when needed.

What can you do?

An interested person could spend a lifetime working and learning about growing their own food, looking after the soil and benefiting from the healthy, green and sustainable life that it creates.

Start small; grow something. A courgette, some mint. Plant a bulb or some sunflower seeds.

Fill pots and tubs, if you have a garden ask for a space to start growing and planting your own food.

Make fertiliser with nettles instead of using toxic chemicals – you can find recipes online.

Contact the Soil Association and grow organic food. Don't add to the earth's problems by using any more chemicals and pesticides.

There is a huge amount of information available about growing your own food.

The Permaculture Association has fascinating books and interconnected narratives on how Permaculture, like Gaia, is somehow all of life connected.

SEE ALSO: Agriculture, p. 27; Climate Emergency, p. 64; Flowers and Plants, p. 153; Gaia, p. 165; Organic Everything!, p. 238; Pesticides, Herbicides and Agri-Chemicals, p. 244

H

HABITATS AND BIODIVERSITY

The demise of our British ecosystems and biodiversity is a terrible loss because so many habitats of species are permanently destroyed.

The State of Nature Report 2019 summarises the fifty top UK conservation organisations' research and their concerns. They show that some 200 animals have declined in numbers by 60 per cent since 1970. The report also brings some small good news: more than 1,500 acres of new wetland were created, including at Steart Marshes in London and Lady Fen at Welney, Norfolk.

There is so much pressure to use land for housing; pressure too for industrial use of water and river systems; the natural forests and woodland continue to be destroyed. These processes are the main reasons that species become extinct, from loss of their habitat.

Filling in wetlands for housing has caused problems. The space between our rivers and solid ground is rich and fertile, with mile after mile of bog, marshes, swamps and fens. To render such sites at the edges of our rivers suitable for building housing requires industrial drainage and destruction of these sensitive biologically rich areas. Everything is drained and scraped away. The earth is cleared ready for concrete pilings and to lay the foundations for our housing estates, infrastructure and factories. Site-sensitive species are lost.

All sorts of mosses, grasses, shrubs and plants that love water thrive here. These marshy areas, filled with the rain of Britain, are the habitat of many creatures, from the tiny water flea daphnia to thousands of species of cold-water fish. Along the banks and in between the marshes are clumps of land, which provide perfect protection and dry bedding for land-based creatures: otters, deer, turtles, newts, damselflies and dragonflies. These rich habitats are a haven for mammals and birds.

This habitat, with its enriched peat soil, is a vital place for marsh violet and sundew. These are the kinds of plants that in turn support both the distinct bird populations and the important invertebrates that are their foods. Creatures without a backbone or skeleton are the most numerous and prolific of all living species: 97 per cent of all species are invertebrates.

The serious threats also come from many environmental sources: climate changes, invasive and non-native species, reduction in river access, urbanisation and pollution, woodland loss and agricultural run-off.

Clear-cutting forests and shrubby areas to create large agricultural fields is a significant problem. Woodland loss in Britain is a widespread concern. Forest Research provide statistics showing woodlands now occupy only 13 per cent of all land with less than a fifth of that dated to 1600 or earlier. Our truly ancient forests are found only in small areas like Epping Forest or London's Highgate Woods. Ancient forests obviously hold a high and complex biodiversity of species, both plant and animal. It is not certain that new woodlands can recreate these species and some are lost for ever. Only thirteen thousand hectares of new woodlands were created during 2018 and 2019.

The long-standing arguments around the planned HS2 rail link highlighted the threat to thirty-five ancient woodlands: some were prematurely destroyed for the new train line while the Covid-19 lockdown was in force despite the presence of nesting birds. Their biodiversity was priceless and an irreplaceable part of our nation's jewels. This kind of species loss has rightly caused disputes between environmentalists and developers.

Changing the flow of a river, creating dams and artificial river areas also contribute to habitat loss, especially when an ancient river loses some or all of its expected incoming water or the water suddenly becomes salty seawater when sea levels rise.

Dr James Robinson, the Director of Conservation at the Wildfowl and Wetlands Trust, says: 'Trying to reverse the declines in our much-loved UK animals and plants is like trying to turn a tanker. But in our wetlands – which are so vulnerable to pollution and degradation – we are creating hope that the direction can be changed.'

Human activity is the main root cause of this destruction, but human beings are also helping to slow the decline. Conservation volunteers number in the thousands. People give up their time to litter-pick, to help with site clearing or give teaching information about nature, to replant native species and shore up crumbling walls, or to plant protective greenery. By protecting and learning about local species and how precious they are, people are making all the difference in a wide number of projects.

It is wonderful to report that despite the wetland, river and woodland species losses, we can now see otters and water voles have returned once again to the rivers of every single county in the UK within the last ten years. During 2019, 138 baby seals were born in the River Thames which caused great rejoicing. Seals are an important indicator species, only surviving in very clean water. The cleaner rivers, wetlands and surrounding environment has helped to protect their native British habitats and we are slowly returning them to their natural state.

What can you do?

Respect and protect your local river, your local wetland, and the home of any creature.

Plant a native tree, support your local woodland or join the Woodland Trust UK. Support charities such as TreeSisters who plant native trees around the world.

Don't drop litter. Don't allow your friends and family to drop litter, especially cigarette butts, which are the most toxic form of litter (see 'Cigarettes', page 61).

Litter-pick, especially around sensitive ecosystems and wildlife areas, to protect and prevent the animals from eating plastics and other unsuitable non-food alternatives.

Join a conservation organisation and make a difference.

SEE ALSO: Climate Emergency, p. 64; Deserts, p. 97; Seas, Oceans and Rivers, p. 278; Soil, p. 283; Trees and Tree Planting, p. 300; Water, p. 317

HAIR PRODUCTS AND HAIR ACCESSORIES

The cocktail of products we are massaging into our scalps can be full of toxic chemicals and animal by-products. There is a huge difference between the standard shampoos on the market and the genuinely eco-friendly products. Most shampoos, especially the cheapest and most widely available, are regularly adding to the toxic chemical overload in our bodies.

Hair products such as shampoos, conditioners, styling gels and dyes contain toxic chemicals such as parabens, sulphates, mineral oils, silicones such as dimethicone, and propylene glycol which is the main active ingredient in antifreeze used for vehicles and fragrances. These can be absorbed through the scalp, exposing our brains and bodies to toxins and pollutants in tiny amounts each time.

Animal by-products can also be found in our hair products, such as keratin, which comes from horns and claws; gelatin, from ligaments and bones; urea, a fancy name for urine; and tallow, which is melted fat.

Washing our hair also uses large amounts of hot water; in fact water-heating is usually one of the main contributors to a household's carbon dioxide emissions. A person who shampoos their hair every day ends up using about 14,000 litres of hot water a year for the purpose, with a carbon dioxide equivalent footprint of at least 500kg. Washing your hair less often helps your pocket and the environment.

Always read the ingredients list on bottles; look for products that clearly advertise that they are free from cruelty to animals, animal by-products or polluting chemicals. Organic and gentle products are widely available.

Hair-dye

According to a survey conducted by The Independent UK, one in twenty women will spend more than £50,000 during their lifetime on products to colour their hair.

Research published in the *International Journal of Cancer* in 2019 showed that the use of permanent hair-dye and chemical hair straighteners was linked to a higher risk of breast cancer in women. The American Cancer Society examined studies which show both a greater risk and no difference in cancer risk.

Hairdressers, however, did have an increased risk of bladder cancer from working with the dyes. Clearly more research needs to be undertaken.

Applying such chemicals to your hair will change its colour but will also damage it, making it weak and brittle; continual use may even result in hair loss. Organic and non-chemical alternative hair-dyes and products are widely available without these problems.

What can you do?

Why not embrace your natural hair colour; it was made for you, it's the best colour for you.

Try to gradually reduce the amount of hair products you are buying and using. Only buy what you need and when you need it, and always look for the organic, sustainable, least packaged options. Consider refill options, which are becoming more and more accessible.

Buy natural, vegan, cruelty-free products – look out for approved logos – and lessen your daily impact on animals.

Search the internet or speak to friends and family about making your own products. A one hundred per cent natural alternative to shampoo is rye flour, and apple vinegar with rosemary and lavender can be used instead of conditioner. YouTube has plenty of do-it-yourself ideas and recipes.

Wash your hair less, and on days when it looks like it needs some attention use a natural, vegan and plastic free dry shampoo from companies such as Primal Suds and Kitenest.

How about not washing your hair at all? Hair begins to make its own natural oils once you stop washing and stripping the hair follicle with soap. Some people report shiny beautiful hair by never washing.

We are seeing the rise of more sustainable hair accessories

such as Tabitha Eve's NObobbles – plastic-free hair ties made entirely from GOTS-certified organic cotton and natural rubber – and Agnes LDN's headbands and scrunchies made from upcycled materials including unwanted clothing. You could of course make your own.

SEE ALSO: Animals and Insects in the Cosmetics Industry, p. 35; Animal Testing, p. 45; Chemicals, p. 54; Cosmetics and Toiletries, p. 80

HAZARDOUS WASTE

Hazardous waste is the poisonous by-product of our industries and our lifestyles. It includes chemicals, batteries, some paints, solvents, pesticide residues, electronic items, inedible oils such as car oil and equipment containing ozone-depleting substances, such as fridges. There is also pharmaceutical waste and clinical waste. Each item has its own special method of disposal. The properties within hazardous waste make it variously an oxidising agent, flammable, irritant, harmful, toxic, carcinogenic, corrosive, infectious, mutagenic or environmentally toxic.

Hazardous waste clearly needs to be disposed of carefully; however, it is so poisonous that it is very difficult to find a way of getting rid of it safely.

According to The World Counts, we produce more than 400 million metric tonnes of hazardous waste each year.

We all have electronic devices in our homes; when they stop working you have to dispose of them responsibly. In order to do this you should take any electronics to your local recycling, repair or refuse centre. Refuse centres have specialist depots tailored to specific hazardous waste.

Electronic products (also known as e-waste), for example,

are either given to charities to repair and sell on or are broken down into their component parts and recycled. Waste industry figures show that of all the electronic products made yearly, only around 17 per cent get recycled at recycling plants annually.

Waste statistics from local authorities show 37 per cent of our household contribution to hazardous waste comes from small items like vacuum cleaners, toasters and kettles and 22 per cent from freezers, cookers and washing machines. Some 23 per cent comes from our disposal of IT equipment including mobile phones, televisions, laptops, computer screens and parts.

Waste is often incinerated, dumped in our environment or in landfill, or traded illegally, which of course wreaks environmental damage. Many people are unaware of both the damage they cause and the illegality of their careless actions.

More recently, richer countries have been paying developing countries to take their hazardous waste. However, developing countries' waste systems are often badly managed and do not have the resources to dispose of mixed and hazardous waste so it ends up being illegally dumped or burned.

Heavy metals such as cadmium, lead and mercury are all found within hazardous waste. When not disposed of correctly they can cause environmental contamination of surrounding waterways and soil, that will persist for a long time into the future. Human exposure to the heavy metals found within hazardous waste can cause serious health problems.

There are evidential reports showing the UK to be the worst offender for hazardous waste dumping in the EU.

Over a two-year period, 2015–2017, a watchdog report conducted by the Basel Action Network (BAN) installed 314 GPS trackers on electronic items disposed of in electronic recycling plants in ten different EU countries. The report stated that eleven tracked items were taken out of their country of origin and were found illegally exported to Ghana, Nigeria, Pakistan, Thailand, Tanzania and Ukraine. BAN estimates that 352,474 metric tonnes of electronic waste is being illegally shipped from the EU to developing countries every year.

Environmental clean-up will take decades, with many nations currently denying any problem at all.

What can you do?

Cut down on buying items that need batteries or that cannot be recharged or recycled and have poisonous ingredients.

Once you have identified your hazardous waste – check with your local council if you are not sure, and look on the product label – see if you can separate it and dispose of it responsibly via appropriate recycling schemes. Electronic goods especially should be recycled carefully.

Read the section 'Heavy Metals', below, for more ideas on how to reduce your own contribution to hazardous waste.

SEE ALSO: Batteries, Rechargers and Instant Energy, p. 49; Chemicals, p. 54; Litter, Waste and Landfill, p. 218; Nuclear Power and Nuclear Weapons, p. 229; Plutonium, p. 259

HEAVY METALS

Heavy metals are found in our everyday items such as batteries and electronic goods. They can be harmful, toxic, carcinogenic, corrosive, mutagenic or environmentally eco-toxic.

Heavy metals are a group of elements within the periodic table that are dense, metallic and, as the name suggests, heavy. They are extracted from the earth and are all around us every day. They are in the water we drink, the products we use and the ground we walk on. Heavy metals are used in industrial processes; they are mined from stable deposits in the earth. This mining invariably causes pollution in the local water table and tiny amounts accumulate easily in lakes and marine waters which has made it a worldwide problem. The metals are processed in factories, and particles can escape in the air and water, posing environmental and health hazards.

Some heavy metals are essential 'trace elements' that we require for our bodies to work effectively. Iron, copper and zinc are needed by humans, in tiny quantities, to help the body function properly and keep us healthy. If you don't have enough iron in the body, for example, you can't make enough red blood cells to carry oxygen from the lungs to the rest of the body.

Too much iron, however, is harmful to our health; and certain heavy metals – for example lead, mercury, arsenic and cadmium, which are commonly found in everyday items such as LCD screens, mobile phones, batteries, refrigerators, dental fillings, plastic buckets and air-conditioning units – are extremely poisonous. They join nickel, zinc and chromium in being the most toxic major pollutants of freshwater systems, rivers and streams because they are so persistent and non-biodegradable. Heavy metals have a high solubility in water and so are easily absorbed by fish and sea vegetables. They accumulate up the food chain and humans eat the fish.

Exposure to high levels of heavy metals is clearly dangerous. Poisoning can occur when there is a build-up of chemicals in the human body, as well as in animals, fish and plants. High exposure in humans to lead can bring about long-term central nervous system damage, kidney problems, blood diseases, stomach pains, headaches and irritability, and in some cases severe mental illness.

At industrial sites manufacturing items containing heavy metals, these elements have been found to leach into the water and air during the production processes unless strict standards and air purification regimes are followed.

Once a product containing heavy metals has been used, it has to be disposed of, and this causes even greater environmental pollution. There are specialist plants where recycling facilities exist to ensure proper disposal of hazardous items containing heavy metals.

New techniques – for example advanced oxidation processes, membrane filtration and chemical precipitation – are being studied, which it is hoped will speed progress towards cleaner processing, manufacture and use of these toxic minerals.

High fish consumption exposes people to potentially dangerous levels of methylmercury now found in swordfish, tuna, shark, bass, pike and walleye. Mercury dental amalgams have also been linked to disease in the body, but medical and academic papers on this subject are too few for firm conclusions. We do know that mercury fillings emit toxic vapours in the mouth, especially during brushing and clenching of the teeth. Many people have removed their mercury fillings and reported better health overall.

Humans are very vulnerable to hazardous chemicals and heavy metals especially when in the womb and in childhood. The best way to reduce your risk is to limit your daily exposure.

What can you do?

Look after your teeth. Mercury is used in amalgam dental fillings unless you specify otherwise.

Avoid nickel-cadmium batteries unless they are the rechargeable type, and always dispose of them carefully.

Cigarette smoking is a major source of personal cadmium inhalation.

Brightly coloured plastics used for household objects like washing-up bowls can contain cadmium. Go for light-coloured plastics and cadmium free brands and dispose of them carefully.

Avoid lead in paint.

Test your water for lead if you are concerned about possible lead in your pipes and your water tank. Check the water pipes coming into your house as many older houses still have lead pipes, which should be updated.

Use a fitted water filter. Run the water to wash the lead out if the taps have been off for a while if you have the old-style leaded pipes or tank.

Some people use iron-rich foods to help absorb and disperse lead accumulation in the body. Suggestions are iron-fortified cereals, spinach, spirulina and other sea vegetables, prunes, beans and lentils and raisins.

Dispose of electronic items responsibly; use the Recycle Now online tool to find your nearest waste recycling point.

SEE ALSO: Chemicals, p. 54; Hazardous Waste, p. 186; Litter, Waste and Landfill, p. 218

HOLIDAYS, TRAVELLING AND RELAXING

We all need a holiday to properly rest, to sleep and relax, and to unwind. We all deserve a holiday.

Where you choose to spend those precious few weeks away is important for you and, of course, the planet. Your choice could have a huge impact on global pollution. In 2018, 126.2 million international air flights taken were by British flyers according to the International Air Transport Association. That's one in twelve of all international airline passengers, making the UK the number one country in the world for people taking non-domestic flights. Research by Airport Watch in early 2020 of men and women aged 20–45 showed that for 50 per cent of all men surveyed who took a flight abroad, their main reason was to attend a short-trip stag event in Europe. One third of the surveyed women who took a flight said their reason was to attend a hen trip for an upcoming wedding.

The number of all flights in Britain on one day peaked at 9,000 in 2018 according to the International Air Transport Association. That's about six planes per minute touching down or lifting off from one of Britain's forty airports.

Passenger numbers were expected to rise by 5 per cent each year. The impact of the coronavirus pandemic on air travel has been brutal, with flight insurance cancelled in many instances, holiday companies going broke, and the numbers of flights for 2020 plummeting while countries instigate stricter quarantine regimes. Some airlines have folded and a once common means of travel has been changed beyond recognition. It is up to us to choose a long-term alternative method of meeting and travelling in the future.

The UN's civil aviation body, ICAO, has introduced a scheme that allows people to 'offset' their carbon cost by buying up carbon credits in the future rather than doing something about their emissions. Other people have been paying to plant trees each time they fly. That cannot last. We can't forward pay our way out of this kind of pollution.

Some research has shown that alternative fuels for the aviation industry are coming and we may well see carbon-neutral flying in the next decade. The Dutch company Urban Crossovers are testing their new jet fuel with a zero-carbon outcome. Willem van Genugten, co-founder and director of the company, has said, 'We want to facilitate synthetic fuel made from CO_2, taken from the air.' Urban Crossovers derives its main component from the air around you – yes, limitless air as a fuel? It does use kerosene, and electricity passed through carbon dioxide and water. At least it doesn't add to the current CO_2 levels in the atmosphere. Kerosene, however . . . what could possibly go wrong?

Before the coronavirus pandemic we were flying across the world as easily as getting on a ferry or a train. All that carbon dioxide was collecting, exponentially growing and causing pollution over every country in the world. On any given day there were 200,000 aeroplanes in the sky.

No one nation can escape the effects as air pollution has no borders. Just take a look at the examples in the table (on the following page), showing the cost of flying as an emission of carbon dioxide from various British airports.

London Heathrow to New York JFK is generally a very busy route, and with thirty flights a day serviced by seven companies and carrying about three million passengers per year, it's been a very lucrative business. In the twelve months to March 2019, according to OAG, British Airways alone generated $1,159,126,794 from this one route (the highest earning from a single route by any airline in the world). Across all the available airlines, over 2.9 million passengers boarded a flight from Heathrow to JFK or JFK to Heathrow in 2018; at 335kg one way that makes nearly a million tonnes of carbon dioxide emissions per year, just from that one flight destination.

Use an online calculator to work out your carbon footprint from all the journeys you have taken by plane. So if you travelled by plane from Edinburgh to Toronto premium/business class, that would generate about 1,226kg of CO_2 and you would need to plant about eleven tree saplings, well maintained, to offset the carbon dioxide created in that one round trip.

CO_2 figures according to the International Civil Aviation Organization (ICAO)

Destination	CO_2 per round trip	No of trees* Economy seat	No of trees* Business Class
London Heathrow to New York, JFK	670 kg CO_2	8	11
Manchester to Disneyland Paris	190kg CO_2	6	7
Edinburgh to Toronto	613kg CO_2	8	11
London to Rome	273kg CO_2	6	7
London to Los Angeles	929kg CO_2	10	15
London to Perth, Australia	997kg CO_2	10	15
Glasgow to Malaga	334kg CO_2	7	9

*Trees you would need to plant to mitigate this amount of CO_2 emissions are shown in the boxes. One tree will absorb one tonne of carbon dioxide over its lifetime. A tree's chance of survival and reaching its forty-year lifespan to fully capture that amount of CO_2 however is not certain, with some trees only having a twenty per cent success rate. You would need therefore to plant five trees to be certain to absorb and capture the CO_2 from one tonne of pollution.

What can you do?

Go the long way . . . If you are taking a gap year and travelling, why would you get on a plane? See the world on foot, by bike, boat, train and bus.

Take your holiday in Britain. Make use of the most sublime and beautiful places in the UK that will bring nature closer to your heart, and make you proud and excited about the wildlife and landscape of your own country.

Save money and go on bike holidays, camping, walking and

rambling or alternative adventures. There are a huge number of specialist holidays with activities in Britain, including bird watching, kayaking, climbing, beach hunting, nature painting, water sports and even working on organic farms in exchange for your holiday with free food.

Walk around one of the large coastal paths, such as the most beautiful Pembrokeshire National Coastal Path or the South West Coast Path, which runs for 630 miles from Minehead in Somerset to Poole Harbour in Dorset. Britain is an island of beaches and coastal wealth; have you explored it?

The National Trust has turned all its holiday cottage experiences into eco-holidays. It now offers electric charging points, solar-powered cottages and many other touches. It is even using 100 per cent of the income to fund its conservation projects.

What about Europe by rail and ferry? There are hundreds of ecologically-themed holidays and events across Europe too. Try nature-based theme parks in Jutland from Naturkraft, which also have Viking museums. Or find an ecological guesthouse in a French seaside chalet or a mountaintop villa or a Spanish wind-powered home.

Ask your workplace to have a sustainable travel plan that allows their employees extra time to travel to meetings abroad on the train, bus or ferry where possible. If you have to travel by plane for work, make sure your workplace pays the environmental cost by getting a reputable firm such as Trees For the Future or the Woodland Trust to plant trees to help mitigate the impact of your flight in the long term.

SEE ALSO: Climate Emergency, p. 64; Festivals, Concerts and Outdoor Events, p. 150; Transport, p. 294; Trees and Tree Planting, p. 300

HORMONES

Our endocrine glands produce vital hormones that control and regulate cells, metabolism and growth and affect our ability to reproduce. This is a slow process, often taking years for hormonal health problems to surface or become diagnosable.

All the glands that make up the endocrine system control many vital bodily functions, including respiration, metabolism and sexual development. They send signals, or messages, to the tissues in the body and give them instructions via the bloodstream. Hormones are essential for all reproductive functions and fundamental to all biological life systems.

The endocrine system comprises of the major glands in the body namely the hypothalamus in the brain, the pituitary, thyroid, parathyroids, adrenals, pineal body, the ovaries in women and the testes in men that all produce these hormones. The endocrine system dictates how much of each hormone is released into the body and when. This of course depends on how much of a particular hormone is already in the bloodstream.

When our bodies are in their natural state, the endocrine system helps us to repair, to reproduce and to grow. We would not be able to grow tall from our childhood if not for these systems and messages. In the last fifty years, however, pollution in the water and the air has been directly affecting human and animal hormones at a cellular level.

In some cases hormones have entered our bodies via the water system. Oestrogens have been found in drinking water unless the water utility use carbon filtration methods to remove them specifically. They are found from women using the contraceptive pill, and more importantly from the run-off from farming manure and animal urine. The animals themselves have often been given hormone treatment and antibiotics, for example so cows increase their milk yield.

In other cases chemicals have been released into the atmosphere in tiny quantities that enter the bloodstream via our breath and directly affect our hormones by disrupting the endocrine messages. Our bodies become stressed and confused with mixed signals and chemical overload. We have added to the hormonal stress in our bodies by overuse of chemicals ourselves at home, via cleaning, washing, cosmetics and even sunscreen.

Our use of chemicals in our homes, in farming and in industry, along with our collective devotion to fossil fuels, has contaminated the air we breathe, the water we drink, the food we eat. As a result we are constantly accumulating traces of these chemicals in our bodies, and over time they affect our health and wellbeing in many ways; in particular they can interfere with the proper functioning of our all-important hormones. The same is true for animals as well as humans.

The chemicals that dangerously affect our endocrine system – collectively termed endocrine disruptors or hormone disruptors – fall into two broad categories: chemicals that change the behaviour of our own hormones and the endocrine glands that produce them; and chemicals that mimic our hormones and effectively act in their place.

Endocrine disruptors may alter the level of a hormone in the body, by making the gland responsible increase or decrease production; they may block a signal; or they may send false or misleading messages of their own.

The end result can be permanent damage to the endocrine system itself, damage to internal organs or to the nervous system, weakening of the immune system, cancers, birth defects, growth disorders, brain development problems and learning disabilities.

Hormone disruptors affect the reproductive system too, and especially fertility. The number of couples presenting with fertility issues is now one in three in some Western countries and there have been serious effects on the quality and quantity of viable male sperm according to the European Environment Agency. Male sperm count is declining in every European country. One important international study released in 2017, and widely reported at the time, found that male fertility levels had reduced by 50 per cent in the past forty years as a result of hormone disruption from chemicals. Disruption in the control of hormone levels affects women's menstrual cycles in particular.

It is thought that these chemicals may be causing fertility problems in wildlife as well, both on land and in the rivers and seas. More studies need to be undertaken of the cumulative effect of this disruption.

We know of around a thousand synthetic chemicals that affect the hormones. Some are actual synthesised hormones, for example the oestrogens manufactured for use in contraceptives and HRT treatments, but the great majority are produced for entirely different purposes, and their hormone-disrupting properties are an unfortunate and dangerous side effect that is only gradually coming to light as we learn more about how our hormones work and about the complex interactions of the cocktail of chemicals we have introduced into our bodies.

From antibacterial gels and hand washes to antiperspirants and toothpaste, our direct exposure to chemicals has soared in recent years. Many of these household products contain chemicals which are generally harmless on their own if used occasionally, but become dangerous with constant and daily use. The European Consumer Organisation, BEUC, urges caution and more protection for consumers. In a major report considering the EU's listing of cosmetics ingredients in 2019, BEUC asked for a tougher and urgent regulatory approach across Europe to the use of endocrine disruptor chemicals in everyday skin products.

Chemicals used in industrial processes find their way into our water systems, rivers and oceans, and get taken up by farm animals, by wildlife, by fish and other aquatic creatures. The herbicides, fungicides and pesticides used in agriculture also contain endocrine disruptors, which can be found on and in the crops on which they are sprayed. The chemicals enter the soil too, and leak from the farmland into streams and rivers, which also puts aquatic and land creatures at risk – some of extinction. Some endocrine disruptors, notably the pesticides DDT and PCBs, have been banned due to their effect on invertebrate animals such as shellfish, and the well-established link with a catastrophic reduction in seal and bird populations.

Humans are then exposed to these chemicals both in their drinking water and in their food: pretty much any everyday non-organic food item will contain traces of one or more hormone disruptors.

As well as pesticides, widely used chemicals shown by strong evidence to be dangerous endocrine disruptors include flame retardants used in electronics and furniture production, phthalates in cosmetics and food packaging, and bisphenols, such as BPA, currently used in the manufacture of aluminium cans (see 'Plastics' on page 253).

Oestrogens in the water, soil and environment from industry and agricultural waste water, essentially urine from animals, have been an emerging concern worldwide. Steroid oestrogens – principally estradiol, estriol and estrone – have been increasing in concentration in the food chain, soil and water. Scientists have pointed to the long-term risk from our intensive agricultural systems, industrial waste through the waterways and the uptake in plants from the soil. In humans there is a link between these hormones and breast cancers.

Much of our processed food contains chemicals called xenoestrogens which are also endocrine disruptors, mimicking oestrogen. Foods wrapped in plastic and cling film contain the highest amount whatever the nature of the food. They are also found in industrially raised meat and dairy products, foods exposed to pesticides, and in parabens found in cosmetics and shampoos. These same chemicals have been linked to breast, testicular and prostate cancers, infertility, miscarriages, diabetes, endometriosis and obesity.

Triclosan and triclocarban are found in antibacterial washes, detergents, deodorants and antiperspirants, liquid soap, shaving products and toothpaste and gels; they can increase the body's resistance to antibiotics.

Industrial processes using chlorine may also unintentionally, as a by-product, create dioxins, carcinogens that have been found in high levels in human breast milk. Dioxins also have a disruptive effect on the endocrine system.

BHA (butylated hydroxyanisole) and BHT (butylated hydroxytoluene) are antioxidants and preservatives commonly found in both cosmetic products and food. BHA and BHT are endocrine disruptors and a human carcinogen according to the American National Toxicology Program. They can cause organ-system toxicity, developmental and reproductive toxicity, cancer and skin irritation. BHA and BHT are widely used in creams and lotions, lip products, hair styling products, make-up, sunscreen, deodorants, antiperspirants, fragrances and perfumes.

Homosalate, which is used in sunscreen and skincare products with SPFs (sun protection factors), works to prevent exposure to the sun's harmful rays. However, homosalate mimics hormonal activity; it is an endocrine disruptor affecting the oestrogen system in particular. It may also enhance the absorption of pesticides in the body.

Octinoxate and OMC, also found in SPF products such as hair colouring, shampoo, lipstick, nail polish and skin creams, are used as a UV filter to extend the longevity of the product when exposed to the sun. Octinoxate has been found to disrupt thyroid function, damage skin cells and cause premature skin ageing. OMC compound is an endocrine disruptor that mimics oestrogen; it has been linked with delayed puberty, lower sperm count, and sperm abnormalities.

The effect on animals and fish of our hormone-disrupting chemicals is still being uncovered. Scientists have found male frogs that have become female due to high levels of atrazine, a herbicide used on corn crops. It can also delay puberty and animals have been found with prostate inflammation because of its use.

Dioxins, lead, arsenic, mercury, perfluorinated chemicals (PFCs) and solvents that include glycol ethers in paints are more chemicals that can affect fertility by damaging hormones. They have been linked to blood abnormalities and lower sperm counts.

What can you do?

The USA-based Environmental Working Group, EWG, has produced a guide to the 2,500 most commonly used household chemicals and how to avoid them, and more detailed guides to some 80,000 cosmetics and personal items.

Women with menstrual problems should try to reduce exposure to all toxins and chemicals. Couples trying for a baby should do the same.

The way to avoid chemical and toxic overload is to actively reduce your exposure to household chemicals, and switch to eating organic. A good diet with plenty of vegetables is key to having a strong immune system. Reduce your dependence on items such as sugar, cigarettes, alcohol and drugs of all kinds which lower immunity.

Wear long-sleeved shirts and hats in the sun; protect yourself without chemicals that block your pores – or go for natural creams.

Make your own cosmetics, toiletries, detergents and food at home so you will be sure of their contents.

Plastics used in drinks and food packaging, including cling film and BPA, can leach toxins into your food and can be avoided. Limit your use of plastics, especially number 3 (phthalates), number 6 (styrene) and number 7 (bisphenols). Look on the bottom of packaging and goods for the plastic identification number.

In particular avoid heating food in plastic containers. Use glass or ceramics. Canned foods may contain BPA in the lining; avoid them. Avoid receipts from shop tills, too.

Be especially frugal and cautious using bleaches, cleaners, laundry liquids and powders, and air fresheners – avoiding completely where possible.

Household paint and solvents can also affect your endocrine system. Always well ventilate a room you are working in and opt for natural paints. Use solvents sparingly.

SEE ALSO: Allergies, p. 33; Animals and Insects in the Cosmetics Industry, p. 35; Chemicals, p. 54; Climate Emergency, p. 64; Species Extinction, p. 285

I

INSECTS, BUGS AND BEES

At a recent Royal Geographical Society of London meeting, the Earthwatch Institute declared that bees are the most important species on the planet, the most important living being. Unfortunately, their celebrity status has come at a dangerous and perilous time for bees. *National Geographic* points out that we have already seen a 90 per cent decline of all bee species.

It's a devastating loss to the busiest pollinator of all flying species. Human beings cannot replace them. Bees pollinate flowers and blossom; they contribute between $235 and $577 billion USD to the annual global food economy according to the Food and Agriculture Organisation (FAO). Bees pollinate 70 per cent of all world food production. Without bees our agricultural system would not survive and humans would starve.

World-famous entomologist Marla Spivak says, 'Anyone who cares about the health of the planet, for now and for generations to come, needs to answer this wake-up call ... fewer bees lead to lower availability and potentially higher prices of fruit and vegetables ... We need good clean food and so do our pollinators.'

What's the cause of the decline in bee numbers? The main

reason is the use of pesticides and herbicides by the agricultural industry. Our farmers spray the fields in order to speed up the growing process and to stop weevils and moths from eating everything. Their chemicals kill everything that is not necessary for the crop to survive. The spray affects not just the soil, but the nearby woodlands, hedgerows and rivers. The objective is to kill all the insects. The soil quality becomes weakened, and it is no longer able to maintain itself and its natural biological system.

Then the farmers have to use more chemicals to try to maintain balance in order for food to grow. It is a dangerous and unhealthy system, for humans as well as the environment.

For bugs, insects and flying beings this is a kind of Armageddon. Everything is sprayed with poison and killed so that perfectly shaped vegetables and fruits can be delivered to your supermarket.

Worldwide, a staggering 41 per cent of insect species have already declined over the last ten years and another 40 per cent face extinction according to the *Journal of Biological Conservation*. This is a serious and critical situation.

Climate change is another factor behind this decline. When unusual weather conditions cause mass extinction events, such as bush fires, typhoons and floods, millions of insects are killed in a short time. Some species never recover and their numbers decline further.

Then think about all the tiny insects you and your friends kill each day. Many people are really frightened by bugs. Despite them being obviously very small, flying insects and spiders in particular cause panic among those humans who are freaked out. People scream and run away. They jump on chairs, they get out implements to swat them or kill them. They spray dangerous pesticide chemicals inside their own homes in order to get rid of them.

Being threatened by insects is usually not a dangerous situation; people need to be calm and assess their encounter more rationally. In the UK there are no native insects (or spiders) that will kill a person unless you are specifically allergic. About twelve people die from allergic reactions to wasp, bee and hornet stings each year. But this fear of insects, spiders, centipedes and other tiny creatures has created a world in which we do not see their collective suffering, or how their extermination has upset the balance of nature. We have undermined them and forgotten how important they are for our actual survival.

Every insect is working hard for Gaia, the ecosystem, Mother Earth, and should be thanked, not squashed to death. Pay attention to your attitude and how people around you speak about and react to insects, spiders, and all the other creatures that get labelled 'mini-beasts'. Make sure you are not part of the problem.

Spiders, for instance, a huge and diverse range of invertebrates, help human beings in many ways. Spider silk is particularly sought after; it is used in medicine, and has been used to heal wounds for thousands of years. Artificially made spider silk is being used to repair skin, blood vessels, tendons and joints. As the science develops, this technology could serve many functions and save many lives.

What can you do?

Don't kill insects or bees: learn to live with them, even love them. Think about and change your attitude to small insects and spiders. How can you be more positive about insects, spiders, bugs and worms?

Get to know one of these creatures: learn about it and find it in nature. Name it and become friends.

Don't kill spiders, ever. They are such an important part of the ecosystem and they keep down the tiny biting and flying insects that get into your home and your sheds and outhouses.

If you come across an insect that frightens you, try not to move. Try not to alarm yourself or the insect. Practise slow, quiet breathing and calmness. The majority will simply buzz away. Most wasp and bee stings happen because people start waving their arms around and screaming and the insect becomes alarmed or frightened for its life. Some bees will die if they sting you; it is their last resort.

Plant insect-friendly flowers everywhere. Send seed bombs as gifts to friends. Don't use pesticides or other dangerous chemicals on your indoor plants or in your garden. Look for natural alternatives.

Go organic in what you eat and the products you use; consider the footprint of your food, make-up, shampoos, cotton and other items. Think of the huge numbers of insects that will be saved!

Organisations that work with insects especially include: Enjoythecountryside.com, Buglife – the Invertebrate Conservation Trust, Action for Insects and the Kent Wildlife Trust.

SEE ALSO: Animals and Insects in the Cosmetics Industry, p. 35; Climate Emergency, p. 64; Cosmetics and Toiletries, p. 80; Organic Everything!, p. 238; Pesticides, Herbicides and Agri-Chemicals, p. 244; Soil, p. 283; Species Extinction, p. 285

IRRADIATION

Irradiation is the 'zapping' of fresh food with radiation to help it last longer and in effect reduce food waste. The technique can also be used to eliminate bacteria that may lead to food poisoning, such as salmonella and E. coli.

Irradiation factories use cobalt-60 or caesium-137 during irradiation. These isotopes emit gamma rays, electron beams or X-rays to irradiate the food. When food is irradiated it absorbs energy. The absorbed energy kills bacteria and delays the ageing of food – for example, by delaying fruit from ripening and vegetables from sprouting. However, irradiation has been found to reduce the vitamin content of food and to affect its balance of essential fatty acids. The appearance and texture of the food changes less during irradiation than in other preservation methods. The irradiation process does not lead to radioactive food.

British law around the irradiation of food states that this process can only be used where it is of benefit to the consumer. A producer wanting to irradiate a foodstuff must prove that the benefits of irradiation outweigh any negative aspects. For example, a benefit of irradiating food is to reduce the risk of foodborne illnesses.

The process of food irradiation is widely criticised by environmental and health groups. They insist that it is unnecessary and potentially dangerous. Environmental groups claim that the process is simply a way of using up nuclear waste and making nuclear radiation more acceptable to the public.

Currently only one place in the UK has an irradiation licence, for dried herbs, seasonings and spices. This is the only group of foods that are allowed to be irradiated and sold in Britain.

The UK Food Standards Agency requires that any food that has been irradiated must be clearly labelled, stating 'irradiation' or 'treated with ionising radiation'. The international symbol indicating food that has been irradiated is the Radura symbol. The Radura is usually green and resembles a plant in a circle.

What can you do?

For more health information contact the campaigning organisation The Food Commission.

Buy organic produce to be sure and look for the green leaf logo on spices and herbs if you wish to avoid irradiated products.

SEE ALSO: Agriculture, p. 27; Fertilisers, p. 147; Gaia, p. 165; Radiation, p. 263

J

JEWELLERY

Our ancestors used jewellery as amulets offering magical protection from perceived hidden dangers, and as a mark of status or rank, as well as adornment. Jewellery was expensive and worn to portray power; it was valued and passed down as part of a wedding dowry or an inheritance. Today's jewellery including watches – from contemporary designs on the high street to fine and unique jewellery consisting of high-quality metals and stones – is enjoyed by most people and has become throwaway and wasteful. This global interest in body adornment has created environmental problems; from the use of nearly extinct animal parts or the mining of rare fine gemstones in luxury items to the widespread use of plastic in cheap imitations, the jewellery market is diverse, complicated and not very eco-friendly.

In 2018 the global jewellery industry was valued at $279 billion; it is predicted to rise to $481 billion by 2025 according to Statista.

Mining

According to Human Rights Watch, around 90 million carats of rough diamonds and 1,600 metric tonnes of gold is mined for jewellery every year, causing serious environmental and social problems.

The environmental impact of mining for gold, silver and diamonds is complex, and damaging to both local people and wildlife. Mining leads to soil erosion, deforestation, polluted waterways and rivers, forced migration and wildlife endangerment.

During the course of mining, if waste products in the form of cyanide and heavy-metal elements such as cadmium, lead, arsenic, selenium and mercury are not carefully contained or controlled, they will pollute the local water and quickly destroy ecosystems, as found in the report 'Dirty Metals' by EarthWorks and Oxfam America (see 'Heavy Metals' on page 188).

Accidents are common. In the Mount Polley mine disaster in Canada in 2014 the gold and copper mine's dam broke and unleashed 4.5 cubic metres of highly toxic residue on to an untouched forest, a nearby lake and land populated mostly by indigenous communities.

Yet toxic waste is also deliberately dumped directly into waterways. In Papua New Guinea, the Lihir gold mine is alleged to dump over 341,000 metric tonnes of toxic waste into the Pacific Ocean over the lifespan of the mine in a report by the *Journal of Rural and Remote Environmental Health* which tracked the impact. Open-pit mining, used when diamonds are near the surface or covered by a thin layer of sand and gravel, leaves particular scarring and problems, but deep-shaft mining is also detrimental to land and neighbouring communities.

The mining boom for gold and silver is heightening the destruction of pristine rainforests and is responsible for poisons entering the Amazon's air and water systems. Deforestation rates in the Peruvian Amazon have increased six-fold due to recent gold mining (2003–2009) according to the *Smithsonian Magazine*.

Threatened Species

The use of animal parts in jewellery is now largely illegal: this includes tortoiseshell from sea turtles and elephant ivory. The importation of coral, including coral jewellery, into the UK is restricted and requires a permit. The Convention on International Trade in Endangered Species of Wild Fauna and Flora (CITES) is working hard to protect these endangered species.

However, illegal poaching of endangered and protected species is still a widespread global problem where the demand for material such as ivory remains high.

Coral reefs support one third of all the fish species in the oceans, but sea corals are at risk of becoming extinct due to human activity, including overfishing and waste dumping.

Freshwater and saltwater pearls are extremely rare. Pearls are naturally produced by oysters as a defence against parasites or irritants. Due to the rarity of natural pearls, the industry synthetically creates the process, which is known as culturing. The culturing process involves surgically opening the oyster shell to insert an irritant. This process is cruel and can cause stress to the oyster.

According to a report by PETA, oyster farming has contributed to the destruction of natural pearl oyster beds by pollution and overharvesting.

Synthetics

Synthetic materials such as acrylic, are used for cheaper high street, 'costume' jewellery. Synthetic jewellery is made for fast fashion purposes and to mimic fine jewellery at a much lower price point, such as synthetic gemstones. Synthetic jewellery simply put is plastic and is derived from the petrochemical industry which is also having a detrimental impact on the environment. See more in 'Plastics' on page 253. Cheap, fast-fashion costume jewellery is commonly paired with low-grade metals such as aluminium, brass and nickel which can cause allergic reactions to the wearer's skin.

However, there is growing interest in sustainable jewellery. Jewellery pieces can be upcycled into new designs which eliminates the need to mine more precious metals. There are also new independent designers looking more closely into the ethics and sourcing processes to upcycle everyday products and even reuse ocean plastics, turning them into new fashion jewellery. The brand Stellen, for example, uses offcuts of bouncy castle plastic, passed on to them from upcycling brand Wyatt & Jack.

What can you do?

Don't buy coral jewellery, it is restricted but items do get through. Refuse to buy ivory, animal skins or trinkets made from any part of an exotic or endangered species.

You might want to join the Environmental Investigation Agency, which exposes trade in animals, or the World Wide Fund for Nature, that campaigns and educates around the world.

Support sustainability; choose second-hand, recycled, upcycled, ethical jewellery brands.

Make your own jewellery.

Brilliant Earth are among the new sustainable companies that recycle old gold jewellery into new items.

SEE ALSO: Fashion and Clothing, p. 139; Heavy Metals, p. 188; Plastics, p. 253; Seas, Oceans and Rivers, p. 278

K

KEEPING WELL AND SAFE

There is a moment many people have experienced – maybe you too – when you realise that many environmental problems are really, really serious. The earth and all its inhabitants are in trouble. It can be a shock. It can make you panic, or feel sad. Some people withdraw and lose hope. Plenty of people are really angry. This is a serious situation and we cannot underestimate the consequences for humanity.

Many people are shocked that politicians and businesses

don't take it more seriously. Many people are angry that the politicians are holding back from calling it a full-blown emergency and putting measures in place immediately to start working towards solutions, especially for the imminent climate emergency and species extinction.

It is also true that many global problems seem so large, so unmanageable and complicated, that you might feel overwhelmed and stressed. Modern environmental emergencies are making people feel anxious.

If you have been part of an environmental emergency in your family you will know the reality of this stress. Thousands of people in Britain and millions across the world know what it is like to lose their homes to events caused by climate change, like flooding, wildfires, drought, heatwaves, tornadoes and typhoons. These extreme weather events increased in number by over 200 per cent and claimed 606,000 lives between 1985 and 1994 according to CRED, the Centre for Research on the Epidemiology of Disasters, and are likely to get worse until we calm down the climate, reduce our carbon emissions massively and start taking responsibility for the environment.

The mental stress of this emergency is real. You don't have to hide your emotions, and you should know that it is perfectly normal to be worried about the environment and the emergency we all face. It is not normal to ignore it.

Ecological stress can make people depressed. By connecting with other people who feel the same, you can help your sadness lift. You can feel hopeful by doing things in your village, your town, your city. By taking action you can start looking towards a greener future. When you start to be in control of what you eat and what you buy, you can be more positive that you are part of the solution. When you make sure that you never drop litter, or kill spiders, or buy unnecessary plastic, you are doing something positive and real for the future.

What is most worrying is that there are people who refuse point-blank to accept any responsibility or consequences from their business or political decisions while claiming to be protectors of the planet. These people are in denial. This is known as cognitive dissonance, the mental conflict a person has when they have two competing inner views about the same subject. This is shown when politicians say they are very concerned about the environment but then continue to increase oil and coal production or give fracking licences without enough environmental standards or community agreement. Eventually this causes mental stress. It would be better if politicians told us the truth and became more truthful to themselves, especially about the climate emergency.

Sometimes politicians are very powerful, making long-term decisions for many people, and sometimes they make really bad choices on our behalf. Citizens need to hold them to account and not accept a green-wash and a downplaying of the problem as if nothing is happening. Otherwise we end up accepting their behaviour and they feel able to continue with their lies, double standards or inaction.

The time for that kind of behaviour is over. We need a mature approach. We need to rethink our lives, our politics, and our economic business model. What has happened with the way we abuse the planet is not good. Don't let people tell you otherwise.

So you are right to be worried, but you should know there are things you can do to help you think straight about the situation, to help calm you down and put it in perspective.

Ultimately this journey is of the soul. It requires you to be awake. To be present and to witness yourself and your behaviour in your everyday life. It requires you to show up and take responsibility for your actions.

Don't be guilty, be active!

What can you do?

Let nature take some of the strain.

Take a walk on the beach, by the river, under a canopy of trees. Lie on the grass and let the sun shine on your body. Be silent with nature and let your worries get cleansed by the water and earth around you.

Keep yourself well. There is no rescuing of the self, other than by yourself. You are loved and valued, so you must remember to love and value yourself.

Forgive yourself. Try to forgive others. They are often struggling in their own way and with their own life and worries. Allowing sadness and forgiveness to flow will ease the stress.

Resolve to do what you can, but not to allow constant worry about what you cannot achieve. When you are genuinely compassionate and generous to your community you will easily know what is needed in the way of support and love.

Laughter and comedy are important. Environmentalists have a difficult job. If you can't laugh it won't be as easy. You don't need to laugh at yourself and the planet; find anything that makes you chuckle and give yourself a daily dose of humour to keep smiling. Fill yourself up with the fun in life.

If you want to give an environmental message and keep your audience enthralled and interested, use humour to inspire and entertain.

Everyone will sometimes need a holiday, will need to find love and support from a family member or a pet or someone else they love. Everyone needs to be nourished, to find rest and relaxation and set aside proper time to reduce stress.

Make time for yourself and rest. Have fun. Have a life. Have friends.

Find love and community. When you meet with other people who feel the same you can support each other.

Some people automatically pretend that traumatic stress is not real. They will hide their pain. This can cause deep psychological problems later. Find someone to talk to. Find someone you trust to speak your truth with. Unburden yourself.

When you have clarity of purpose you will feel better about the world. Work out what you feel positive about, what matters to you. Work out what you are inspired by. Do it. This will feed you, nurture you and keep you sane.

Make a personal, spiritual connection to nature. Listen to the earth. She will tell you what to do. Your connection to the earth is your path, your lifeline to the most potent, stunning, awesome, wise, incredible form of energy.

SEE ALSO: Climate Emergency, p. 64; Environmental Stewardship, p. 133; Ethical Living, p. 135; Gaia, p. 165

L

LITTER, WASTE AND LANDFILL

> 'There is no such thing as "away". When we throw anything away it must go somewhere.'
>
> Annie Leonard, *The Story of Stuff*

This year, you will probably throw out ten times your own body weight in waste. The waste doesn't go far away. Until the past fifty years we mostly burnt or buried our rubbish in the ground.

Textiles, paper, glass, plastics, metal, toxic substances and kitchen waste are put out on our doorsteps – British councils calculated about 593kg per person per year in the UK during 2019.

In Britain our waste disposal is managed on an ad hoc basis in each local authority area. It is usually taken to a local council-run landfill site, with 50 million tonnes of industrial, commercial and domestic waste sent there every year, according to government statistics. Recycling is the most well-known system to divert waste from landfill (see 'Recycling' on page 265). The waste hierarchy however asks us to *refuse* first, by not purchasing, then *repairing, reusing* and *re-circulating* before going further down the waste chain including *recycling*. Anything else that doesn't fit into these categories ends up in landfill with their useful energy wasted.

In the UK we are currently facing the threat of more than 1,200 coastline landfill sites, holding decades of waste, becoming dangerously exposed due to coastline erosion and rising sea levels. This will cause toxic, previously hidden waste to spill out, polluting seas and beaches. This threat also poses health risks to local communities, primarily from the contaminated waste reaching the water table. Research conducted by Queen Mary University examining a landfill site on the east coast of England concluded that the area was a 'significant contamination site'.

An alternative way to dispose of household waste is to incinerate it, and the idea of producing energy from this process seems attractive. But there are long-standing problems. If a mixture of waste is burned, toxic gases are produced. Think of the acrid smell that comes from many household bonfires: it comes from burning plastics. If the temperature in the incinerator is not high enough – it needs to be over 900°C – plastics, pesticides and wood preservatives give off poisonous dioxins and other chemicals in the air, adding to air pollution.

These and other poisons remain in the ash, which is then sent to waste sites for burial. Incinerating mixed rubbish produces more pollutants per unit of energy generated than any other kind of fuel.

Many countries with incineration plants that started early in the technology's history have found they do not have enough raw waste now and are importing waste from elsewhere to fulfil the capacity required to run them to be economically viable. In Sweden recycling and waste-reduction programmes have been so successful the country is approaching zero waste but still has 33 waste-to-energy WTE plants in operation.

In 2017 in Britain 45.7 per cent of household waste was recycled, while 2.4 per cent was sent to landfill according to government statistics. In 2018 Jenny Jones from the UK Green Party released a report that challenged the economic viability of incineration. In a statement at the time she said, 'There is a logic to generating energy from the waste that we cannot recycle or reuse, but it is meant to be the last resort option. What we have created instead is a market-driven system of incinerators which constantly need to be fed.'

But our waste isn't always put through a council system. Careless and illegal waste disposal – from littering or putting hazardous-waste items in with the general waste, to fly-tipping and illegal dumping – is a big contributor to environmental damage, adding to the pollution in our seas as well as marring our countryside and city streets.

With fly-tipping reaching a ten-year high in 2019, environmental charity Keep Britain Tidy blamed confusion over complex disposal systems introduced by local authorities. In 2017/18 fly-tipping was up eight per cent on the previous year, costing local authorities £129 million to keep our streets clean.

Cases of fly-tipping range from small-scale dumping of black bag rubbish to large-scale professional operations. A 2019 incident in Stockport saw a heavy goods vehicle dump a 50ft-long pile of rubbish on a public cycle path overnight. Professional fly-tippers are scammers who get paid by an individual or a company to responsibly dispose of bulky items, but instead they illegally dump the rubbish and leave with their pockets full. During the coronavirus pandemic lockdown the waste and recycling centres were closed and fly-tipping increased in the UK to an all-time high, causing damage which may take years to rectify.

No matter how big or small, fly-tipping is an unsightly and costly offence, and a problem that needs to be tackled. Local authorities should impose tougher penalties on those found fly-tipping. The manufacturers of bulky goods must start accepting back old products for recycling.

The dumping of waste is not just a British problem: it is equally as serious, if not worse, in developing countries. Western nations, including the UK, Canada and Germany, have been sending unwanted rubbish and recyclables overseas in an effort to reach recycling targets and reduce their own landfills. These consignments have included mixed electronic waste containing heavy metals. The receiving countries, such as Malaysia, Indonesia and Vietnam, accepted this rubbish as a source of income, but with contaminated waste and hard to recycle items piling up it has created toxic wastelands. In 2019 the Malaysian government started an investigation into Western shipments and have since begun sending back waste to the countries it originated from.

Waste and litter are polluting our rivers, lakes and seas and killing wildlife. Waste can find its way into our waters through street and beach littering, carried by the wind and rain into our drainage networks or rivers that then flow into the sea. Major rivers around the world carry an estimated 1.15–2.41 million tonnes of plastic into the sea every year – that's up to 100,000 rubbish truck loads according to Greenpeace.

An estimated 8 million metric tonnes of plastic waste enters the oceans every year as reported in *Science* magazine. Dumping the waste produced by one environment into another is not an answer.

What can you do?

The sea starts here! Make a note of your nearest drain in the street outside. This is where the sea begins for you. Respect it; don't allow litter to enter it. #TheSeaStartsHere

If you see fly-tipping, report it immediately to your local council. Don't let wildlife eat it.

Sort your own rubbish carefully. Recycle and reduce your waste. Use all your local council services. Look online: use recyclenow.com.

Support any schemes or groups in your area that try to clean up waste and litter. Set up a local litter-pick community to raise awareness of street litter. Join the CPRE annual Green Clean.

If you can't find a safe way to dispose of a particular product, send it back to the manufacturers, with a letter asking them to dispose of it for you.

If you have a garden, start a compost heap from your kitchen waste to get free mulch for your vegetables and flowers. That could be up to 35 per cent of your waste problem solved.

You can cut down on the waste you produce by not buying so much in the first place.

SEE ALSO: Consumption and Consumerism, p. 77; Hazardous Waste, p. 186; Plastics, p. 253; Recycling, p. 265; Seas, Oceans and Rivers, p. 278; Water, p. 317

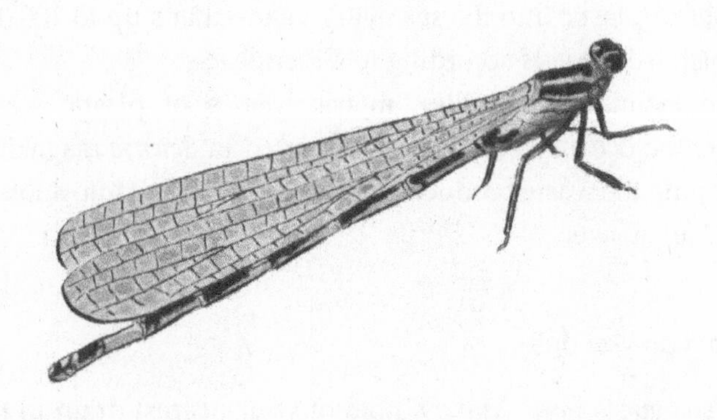

M

MALNUTRITION

Today, nearly one in three people in the world suffer from malnutrition as a result of not eating enough food or not eating the right foods according to the UN. More than two thirds of these people live in Asia, with the majority of the rest in Africa and Latin America.

With the world's population estimated to rise to ten billion by 2050, and the threats posed by the overuse of the world's agriculture, we are at serious risk of more people becoming affected by malnutrition, starvation and even death.

One effect of the coronavirus pandemic was that more than 260 million people quickly faced acute food insecurity during 2020, and organisations and charities seeking to offer humanitarian relief, including the United Nations, have been pushed to their limits.

In theory, we should be able to produce enough food for the whole planet to eat enough, but resources and distribution are being mismanaged. Environmental concerns add to the burden. The overproduction of meat has resulted in valuable crops that could feed people being used instead to feed cattle.

One country's demand for its favourite fruits and exotic nuts is causing another to suffer soil erosion and the drying up of fresh water supplies and wells.

According to the charity UNICEF, globally approximately 3.1 million children die each year from malnutrition alone. In addition, malnutrition and hunger are also contributing factors to more than half of all global child deaths as this can make children more vulnerable to illnesses. The 2018 UN Global Nutrition Report shows that a staggering 150.8 million children have stunted growth because they are not eating enough calories each day. That's 22.2 per cent of all children on the planet.

Living in a first-world country does not make people exempt from malnutrition. In 2016, almost three million people in the UK were estimated either to be living with malnutrition or to be at risk because they do not eat enough according to the national think tank, the Food Foundation. In the same year, malnutrition was listed on more than 350 death certificates in England and Wales and an estimated 1.3 million people over the age of sixty-five were said to be suffering from malnutrition.

In 2019 there were more than 2,000 food banks in the UK according to the Independent Food Aid Network. Over 1,200 of these food banks were run by the Trussell Trust charity. The Trussell Trust has identified a sharp increase in the number of households in need in Britain over the past five years. The use of food banks run by the charity grew by 73 per cent between 2014 and 2019.

Before the coronavirus shutdowns and the subsequent economic downturn, they reported the top three reasons for people visiting food banks to be: incomes not covering essential food costs (33.11 per cent), delays in payment of benefits (20.34 per cent) and the impact of changes of benefits (17.36 per cent). The coronavirus pandemic emergency meant a record in demand was matched with a depleted and under-resourced supply. In a time of rising and sudden unemployment and a change of eating habits with the closure of many restaurants, food banks are a fundamental and critical source of food for many families.

While many remain hungry, the distribution of food is a critical problem.

Globally it is estimated that in today's society we are wasting some 1.3 billion tonnes of food according to the Food and Agriculture Organization of the United Nations. The UK is the worst offender in Europe. If all of the wasted food across the US and Europe were gathered together, over a billion hungry people could be fed.

What can you do?

Educate yourself. You can find out so much about world hunger with a quick online search, or just by reading the newspapers.

WRAP has an online food information factsheet, 'Food surplus and waste in the UK – key facts', January 2020 (wrap.org.uk).

Help others who might be suffering silently around you. If you can, support a local food bank by donating food to those in desperate need. Don't waste your food.

Support projects and ideas that increase the amount of free food available in your community, such as fruit tree planting and vegetable patches, guerrilla gardening and shared allotment programmes.

SEE ALSO: Agriculture, p. 27; Climate Emergency, p. 64; Environmental Stewardship, p. 133; Food Waste, p. 156; Soil, p. 283

N

NOISE POLLUTION

Noise pollution surrounds and affects us all day and every day. It is an invisible problem. Many people are not aware of the damage noise can cause.

From road traffic, music concerts, industrial machinery and aeroplanes to noise pollution from ships, sonic booms and human activity, all these create a high level of inescapable sound in our environment.

When noise becomes too loud, persistent and unwanted, it poses a threat of ill health to all animals and birds, creatures in the soil and humans.

People who live near motorways and airports have been found to visit their doctors more often. They have more stress-related illnesses and more psychological problems. Loud sounds destroy sensory cells in the inner ear, and these do not heal, which can eventually lead to deafness. Exposure to persistent or loud noise can also cause high blood pressure, heart disease, sleep disturbances and stress. These health problems affect all age groups of people but most especially children.

Be aware of your surroundings: playing music too loud,

having the volume of the TV too high and speaking too loudly all contribute to noise pollution. You do not need to walk around whispering, just be courteous to those around you, including nature.

Noise pollution is extremely disturbing for animals and wildlife. Animal hearing is a lot more sensitive than humans'. They can use their acute sense of hearing to navigate, find food, attract mates and avoid predators. Noise interference makes this very difficult for them, which can affect their ability to survive.

Noise and sonic pollution in the ocean is also very disruptive to sea creatures. Noise from ships, shipping vessels, oil drills, mining, sonar devices and human activity are all causing what should be a calm environment to be very chaotic for all marine life. Just like their contemporaries on land, sea creatures rely on their sense of hearing to communicate, feed, find a mate and navigate. High levels of noise pollution make this increasingly difficult for them.

A completely unnecessary source of noise pollution comes from fireworks, and this affects both humans and animals. In humans noise pollution from fireworks can cause hearing loss and tinnitus, an inner ear condition. The effects of fireworks can be just as devastating to animals, leading to phobias, panic attacks, hearing loss, hyper-vigilance and even death from the fright.

Many owners report their cats and dogs being terrified of firework displays. When the fireworks are near trees and bushes, the noise wakes up sleeping birds and mammals and their young. The terrific noise gives them a fright, panicking them and alerting their little hearts to danger and flooding their bodies with adrenalin.

What can you do?

We can choose what we listen to, for the most part. We can choose to consciously walk in nature and listen to the sounds of nature. This is a beautiful form of healing. There are studies that conclude that the sounds of nature – a running river, leaves rustling in the trees, the ocean roaring, birds singing and chirping at sunrise, even silence itself – cause inner peace to arise in the senses and stimulate endorphins. A person feels happier, calmer and their blood pressure drops. Listening to nature can heal you.

Are you aware of what the noises are around you at home? Have you chosen them? What kind of music or television or radio can you hear? Is it on consciously or just as noise wallpaper? Do you cause pain and annoyance to others around you because of your choices?

How do you communicate with those around you? Are you clear and confident? Do you sing with joy? Or murmur with passive-aggressive rage under your breath at every chance? Are you too loud? Do you shy away from saying anything? How does the sound of your voice communicate your needs and truths to those you love? Be aware of the noise you and others create. Is it necessary?

Try to avoid extremely loud noise. Don't stand too near the loudspeakers at music concerts. Use earplugs for all festivals and especially for long-term festival and event participation.

If you have noisy neighbours, negotiate a 'noise agreement' with the people who live around you. If they want to play music late at night but you have to get up early, see if you can come to an arrangement that they can only make noise at the weekends, when you can sleep in – and vice versa! If not, your local council will have a Noise Abatement officer who can help to resolve very loud and repeated noises.

Don't support firework shows unless they are silent ones. Write to your local council about their community events that have fireworks. Support and campaign for non-invasive light shows to celebrate special occasions instead.

If you are suffering from continuous loud industrial noise, get in touch with the Noise Abatement Society, which has a number of suggestions as to how to deal with it. They also campaign against disruptive noise generally.

SEE ALSO: Gaia, p. 165; Keeping Well and Safe, p. 213

NUCLEAR POWER AND NUCLEAR WEAPONS

Nuclear power is one of the costliest forms of energy available today. At the moment it provides about 10 per cent of the world's commercial energy, from around 450 nuclear power reactors.

When nuclear power first began to be exploited, scientists claimed the electricity would be so cheap that it would be given away free. Instead, the nuclear power industry costs society a great deal of money and creates a wide range of environmental, social and political problems. It is a very controversial issue.

Nuclear power is generated from the naturally radioactive element uranium, which is mined in the USA, Canada, southern Africa and Australia. Health and safety facilities at the mines are mixed. In southern Africa there have been worries that staff involved in mining uranium have an increased risk of contracting cancer. Uranium is classified as a human carcinogen, although in its inert state in rock it is not harmful.

The 'fuel' that drives a nuclear power station is a rod of uranium inserted into the reactor in the station's heavily shielded concrete core. The reactor operates like a very slow, controlled nuclear bomb, generating huge amounts of heat. In the type of reactor most commonly installed, this heat is used to boil water into steam, which in turn is used to produce electricity.

Although many nuclear power stations in the UK are closing, a new generation are proposed in Somerset by EDF Energy at the Hinkley Power Station site. The costs have already spiralled to between £21 billion and £22.5 billion and it is expected to be on stream by 2025 and supply around seven per cent of the UK's energy needs. The price, however, is rising and already it is far more expensive than renewable energy schemes. Greenpeace UK's chief scientist said: 'The new Hinkley nuclear plant looked like a bad idea when it was first proposed, and it's got worse ever since. New offshore wind now costs less than half as much as Hinkley, and it might get even cheaper by the time the much-delayed reactors crank into action. An energy system run on renewables would be the cheapest, fastest and most reliable way to cut carbon emissions.'

Nuclear power stations are not 100 per cent safe. There have been accidents such as that in Chernobyl in Russia in 1986, and Three Mile Island in Pennsylvania in the United States in 1979. It is very costly to clean up a nuclear accident.

In 2011, when an earthquake and a 15-metre-high tsunami hit a nuclear power plant in Fukushima, Japan, the reactor exploded: an accident that caused death and environmental destruction. The radioactive clean-up has not yet finished.

Contaminated water from the accident has leaked into the Pacific Ocean. There is already a million tonnes of contaminated water in barrels on the site and the Japanese government has said it may need to 'dispose' of that in the Pacific Ocean as well. The idea is that the ocean water would dilute the radioactive water from the plant. Local people and the fishing industry are understandably concerned.

The accident was so damaging to the local environment that

fish caught there can no longer be eaten. The plant has not been able to reduce the radioactivity around the site to a level that is safe for local people to return. Between 200,000 and 600,000 people were evacuated according to various statistics from the government and nuclear industry. It is estimated that only half of those who left have returned, even where the radiation levels are dropping.

People have died from the trauma of the accident and many will have suffered long-term health problems that have yet to reveal themselves.

Nuclear power production is linked to the manufacture of nuclear arms as plutonium is used in weapons of mass destruction.

The British government has a stockpile, referred to as a 'strategic arsenal', of 120 active nuclear warheads with a 13,000km missile range. The government says the UK needs its own independent nuclear deterrent in order to stop other countries from having and using nuclear weapons.

Nuclear weapons such as the Trident missile, which is stored in Scotland, are deadly, designed to kill a vast number of people and destroy an area completely. They are the most dangerous weapons on earth.

Politicians repeatedly say that they don't want to use nuclear weapons, but the world has already witnessed them being used in war. In the Second World War the Americans dropped atomic bombs on the Japanese cities of Hiroshima and Nagasaki, killing 28 per cent of the population. About 220,000 people were immediately injured and killed; many more suffered long-term health effects. Even people far from the bombsite found they were vulnerable to long-term health problems, such as cancers, and having babies with deformities.

The effect on the government and people of Japan was profound. They looked at themselves and their reasons for war.

Japan declared itself against using this type of bomb and its government has signed many treaties over the years prohibiting the use of nuclear weapons and calling for peace and peaceful resolutions to war.

Japan has a weapons policy that means it adheres to the three principles of non-possession of nuclear weapons, non-production of nuclear weapons and non-introduction of nuclear weapons. Japan is the only country in the world where nuclear bombs have been used.

The spread of nuclear weapons and materials to make bombs has worried the international community. They know that nuclear waste and fissile material can be shared with dangerous armed groups and terrorists. The expansion and development of nuclear power across the planet has increased the risks and the movement of fissile material. Plutonium, another radioactive element that can be used to create a nuclear bomb, is a dangerous by-product of the nuclear power industry.

What can you do?

Save electricity. Energy efficiency reduces the need for more dangerous methods of energy production. Using less energy ultimately means that there will be less demand for nuclear power stations to be built.

Carefully using energy and electricity saves you money but it's also good for the planet in many ways. Reduce your need for large-scale energy production: instead buy your electricity from a supplier such as Green Energy UK that is 100 per cent renewable.

Join a campaign against nuclear power and weapons, such as those run by Campaign for Nuclear Disarmament (CND), Friends of the Earth and Greenpeace.

Campaign for peaceful communication and an end to war. Learn how to be peaceful in your life and relationships, ask for peace. Envision peace and work for peace. Peaceful people and communities produce peaceful governments and countries that do not need to use nuclear weapons.

SEE ALSO: Climate Emergency, p. 64; The Development of Humanity, p. 102; Energy Efficiency and Energy Conservation, p. 116; Plutonium, p. 259; Radiation, p. 263

ONE PLANET LIVING

When you are aware of your lifestyle and the effect it has on the planet you will realise that it isn't one issue – just cycling instead of driving a car or just eating organic – that matters. Your life and overall consumer purchases will make a difference.

One Planet Living is a vision of humanity living within the earth's resources, not wasting, nor spoiling what we have, rather developing sustainably. Many green organisations have developed a computer model where you can look at your lifestyle and reduce your own consumption day to day. The One Planet Living campaign has been developed by the Bio-Regional Group and aims to inspire people to live within the earth's true capacity.

At present, the average person in the West uses the equivalent of around three or four planets' worth of resources according to the campaign. This means that people living in rich countries are using up the planet's resources as if they have three or four planet earths instead of the one we actually have. There are a finite number of resources on the planet, and without a concerted effort to protect them and ensure they are sustainable for people to enjoy now and for the next generation, they are likely to be destroyed, polluted and lost.

Some very wealthy people are using even more resources, and the lifestyles of those who take regular first-class flights, live in two or more homes, have several cars and eat out five or more times a week take the footprint statistics to a new high. This super-wealthy lifestyle, without any environmental mitigation, would mean that if everyone on earth did the same we would need 20 or more planets for each person! This is worrying because we clearly cannot continue to develop in this way. Our vision of success and wealth must change as part of our ethical-sustainable future.

The rich Western countries use more than their fair share of what there is and the most wealthy even more. It is obvious that we do not have more than one planet! We must be more aware and cleverer in our use of raw resources so we help to replenish them, not destroy them, and share them with the rest of the earth's inhabitants.

Affluence and overconsumption is killing the planet. A recent study for the journal *Nature Communications*, told us that only 0.54 per cent of the world's population – just 40 million people – are responsible for an extraordinary 14 per cent of all carbon equivalent pollution and the use of resources that equal that.

One Planet Living is a goal we must really embrace. Millions of people already do. Our consumption of products, clothing, cars and fast food is contributing to the climate emergency. Our throwaway consumption is creating the plastics disaster in the oceans.

The Bio-Regional Group's ideal of One Planet Living is guided by what they call their 10 principles that will make a sustainable world possible. These are discussed in detail in their free online-community help packs covering the key issues for people to work on: 'Health and happiness'; 'Equity and local economy'; 'Culture and community'; 'Land and nature'; 'Sustainable water'; 'Local and sustainable food'; 'Travel and transport'; 'Materials and products'; 'Zero waste' and 'Zero carbon energy'.

#OnePlanetLiving

What can you do?

We live within one beautiful, stunning, awesome planet called Earth. All beings who are alive are valued and loved by someone. All alive are worthy of clean water, fresh food and to find a place, a role, and reach their potential.

The ones who have so much might not be greedy. They might have grown up being showered with love by their parents and have received a multitude of gifts and presents all their lives. The ones who have so much might not be unconscious or unkind either. They might have grown up as bright people, yet struggling in their own way.

Each person is on a journey of discovery. When their discovery allows them to be awake and conscious to the planet's needs, and the needs of the other people around them too, then we hope humankind will be just that: kind.

Generosity and kindness, with a humble manner, will help the transition. This new way of thinking and living will not need to find power or status in labels, brands or logos.

Where are you on this journey?

Don't fly. A long-haul flight contributes around half of an individual's yearly carbon footprint. Take the train, take a bus or a boat, walk or cycle. Say no to flights that are not urgent. Take your holiday the long way round; make your journey part of your holiday instead.

Reduce your environmental impact by avoiding meat, especially red meat, and dairy products, to make a significant impact on your carbon footprint.

Consume less, buy less overall, wear your clothes and repair them, refashion them. Buying more things won't make you happy; plenty of studies have shown this. Invest in yourself, empower yourself.

Buy local, support local artists, cooks, butchers, etc.; buy fairly traded and ethical goods.

Don't buy stuff that adds to pollution or gets thrown away quickly.

Be aware of your energy use. Turn off lights, your TV, unnecessary heaters, devices and computers.

Stay active; enjoy nature and being in nature as an alternative to shopping.

Live a zero-waste lifestyle. Reduce your waste trail and reduce your carbon footprint.

Repair things when they break, take a bag with you when shopping, use a reusable water bottle. Most of these ideas are common sense, ecological common sense.

The Global Footprint Network has developed a calculator so that you can properly look at your ecological footprint. It can tell you how many planets you'd need if everyone lived like you do. It makes for grim reading.

For more information you could look at the One Planet Living computer app from the Bio-Regional Group, and similar computer-aided questionnaires from World Wide Fund for Nature.

Explore *BBC Good Food*'s Seasonality Table, and the campaign 'Love Food Hate Waste'.

SEE ALSO: Climate Emergency, p. 64; Consumption and Consumerism, p. 77; Fossil Fuels, p. 160; Your Carbon Footprint, p. 322

ORGANIC EVERYTHING!

Organic products can include food, cosmetics, essential oils derived from plants, textiles and cleaning products.

Simply put, organic is non-toxic, non-chemical, environmentally, socially and economically sustainable production.

Organic production pays close attention to the best interest of the environment. Organic means more wildlife, biodiversity, not using antibiotics in rearing livestock and as a result avoiding genetic modification.

Organic methods aim to retain the long-term fertility of soil. They avoid artificial fertilisers, pesticides, regulators and additives, and, of course, irradiation. The use of genetically modified organisms (GMOs), or products produced from or by GMOs, is generally prohibited by organic legislation, too.

An organic farmer will employ some of the oldest principles of agriculture, keeping the soil fertile by using soil-improving plants like clover, and natural fertilisers like animal and plant manures, instead of chemicals. Instead of growing just one crop, organic farming rotates crops, growing something different every year for a three- or four-year cycle so that the soil does not become drained of nutrients. Keeping things varied in this way also means that pests find it harder to attack the crops, because the crop keeps changing and they have to adapt. Organic agriculture also requires that animals are kept in natural, free conditions.

For products to be labelled as certified organic at least 95 per cent of ingredients must be organically produced. EU-wide rules require organic products to be approved by an organic certification body, such as the Soil Association or the Euro Leaf, which carries out regular inspections to ensure the ingredients meet a strict set of criteria.

The price of organically grown products reflects the work

involved in ensuring their purity. Organic is generally more expensive, but as demand for organic products grows, more farmers will see both the business and the ethical potential for turning their land over to organic production. We hope that as more farmers start to produce organic, the prices will come down.

What can you do?

Look for organic everything: organic textiles, beauty products and food. Ideally, anything that goes in or on your body should be as organic as possible.

With organic food, many people say it tastes better than mass-produced food. If you are on a budget and cannot afford to go completely organic just take it slowly. Prioritise the most important things.

Buy organic sometimes at least, even if you can't buy it every time. See if your favourite brands offer organic alternatives.

Grow your own organic food in your garden. Follow the Permaculture philosophy of nurturing the soil.

SEE ALSO: Chemicals, p. 54; Climate Emergency, p. 64; Grow Your Own Food, p. 177; Your Carbon Footprint, p. 322

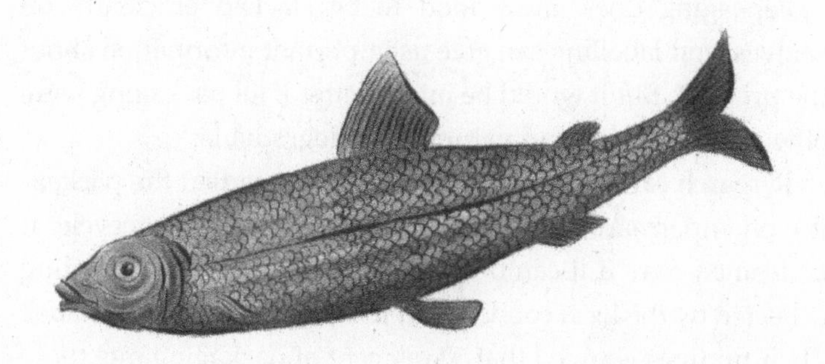

P

PACKAGING

Some 80 per cent of our household waste is packaging: plastics, glass, paper, cardboard, tins and non-recyclable mixed materials. Some items we purchase contain more packaging than product and it's not always recyclable.

Nature has provided its own wrapping for many foods. Oranges, kiwis, nuts and vegetables such as a simple potato, a cabbage or a cauliflower – all come in their own packaging, and yet we waste a huge amount of resources wrapping them up again. Almost half of packaging used by the major UK supermarkets cannot be recycled easily, and so it ends up dumped in our overloaded landfill sites.

Packaging does allow food to be stacked efficiently on shelves, and labelling can give us important information about the product. But it would be much better if all packaging were efficient, recyclable and naturally biodegradable.

Research carried out by *Which?* in 2019 studied the packaging on supermarket products for how easy it is to recycle. It is deemed easy if it can be placed in the household recycling collected by the local council, and it says so clearly on the label. Their findings showed that 52 per cent of packaging met these requirements – including pieces with cardboard, glass and plastic. But 42 per cent of the total supermarket packaging was found to be labelled either incorrectly or not at all.

One of the biggest villains is the plastic bag, which we take away in their millions from shops and supermarkets. The vast majority of these cause long-term problems for the planet. Since 2015 large retailers in England have been required by law to charge for all plastic carrier bags, including the single-use bags they used to give away; however, these large retailers are still providing small bags suitable for loose fruit and veg and takeaway foods for free, and many shoppers are choosing to use them, and to pay for carriers. According to Greenpeace research, sales of the 'bag for life' had risen to £1.5 billion in 2018. The research also found that overall supermarket plastic use has risen to more than 900,000 tonnes in 2018, despite pledges by retailers to cut down.

An investigation by the *Guardian* found that plastic bags can contain tallow – an animal by-product commonly derived from cow's fat. It is used as a 'slip agent' to keep the bags from sticking together. Plastic bags are therefore not suitable for anyone practising vegetarianism or veganism or those who have strict religious beliefs.

Plastic carrier bags are banned in many parts of Asia and Africa, and there are partial bans in other parts of the world.

What can you do?

Try to avoid over-packaged goods when shopping. Look for unpackaged alternatives; support local independents selling unpackaged food, like a local fruit and veg stall, a farmers' market or a zero-waste bulk store where you can refill your own containers.

When shopping at large retailers, avoid packaged foods and opt for loose fruit and veg: vote with your wallet.

Try to avoid packaging that is made of mixed materials, like plastic-backed paper or foil-backed plastic. This makes the packaging impossible to recycle. Recycle all recyclable materials.

Refuse plastic carrier bags and use your own reusable shopping bags.

SEE ALSO: Disposable Products, p. 105; Gifts, Presents and Greetings Cards, p. 167; Litter, Waste and Landfill, p. 218

PAPER

Paper and cardboard is a natural renewable resource and important material used worldwide. In 2017 according to Statista approximately 419.7 million metric tonnes of paper and cardboard was produced globally.

In the UK alone 12.5 million tonnes of paper and cardboard is used every year according to industry and council waste statistics. This equates to a fifth of all the waste produced in the UK annually. Of this, 80 per cent gets recycled, but the remaining 20 per cent may end up in landfill sites. This recycling rate is so high because campaigners and businesses joined together and created a successful market for recycled paper goods. Global paper recycling is at around 58 per cent.

By wasting and throwing away paper we are helping to destroy the world's woodlands, forests and rainforests and threatening the balance of life on the planet. The paper hierarchy promoted by environmental campaigners advocates waste prevention as the first goal, and then reuse, recycling and energy recovery, and all of those before throwing it away in landfill.

To meet our paper demands native British and European forests are being replaced by thousands of acres of conifer plantations. These huge 'artificial forests' are often planted in soil that is not meant for this type of tree, and the trees are forced to grow at speed. This leads to a loss of soil nutrients, plants, wildlife and habitat in the area. Monoculture trees grown from one species used only for industrial paper manufacture ensures little biodiversity in the forest and has caused an irreplaceable loss of native species. Large numbers of these trees suffer from illness such as 'windthrow' when they have insufficient roots and fall over on a windy day. In some Scottish monoculture plantations forty per cent of the trees have been lost or damaged in this way.

Rainforests in Central America and Asia are being cut down and replaced with eucalyptus trees to be used for paper production. The eucalyptuses soak up so much water from the soil that they affect the delicate water balance. This causes devastating soil erosion and environmental damage.

Because trees soak up carbon dioxide, we are increasing the quantity of carbon dioxide in the air when we cut them down. Overuse of paper and wood products exacerbates the climate emergency.

Worldwide, four billion trees are cut down every year for paper and the World Bank estimates that 3.9 million square miles of forest are lost each year.

It makes ecological and economic sense to recycle all paper and card. It takes 70 per cent less energy to recycle paper than it does to make it from raw materials. Recycling paper creates 73 per cent less air pollution than making new paper from raw materials. By doing this, we would help preserve the native forests. By recycling just one tonne of paper, we can save seventeen trees and 4,000 kilowatts of energy, which is enough to power a house for a year.

What can you do?

Reuse all the paper you can. Use the back of letters and envelopes for scrap paper, or to make a new notebook. Reuse paper bags from vegetables for sandwiches and snacks, don't switch paper for disposable plastic. Give old cereal boxes to your local playgroup so that they can use the cardboard for making models. Open envelopes carefully so that you can use them again.

Use recycled writing paper and toilet paper, which you can buy at many stationers and supermarkets. Urge any groups you work with to use recycled paper, too.

Try to avoid buying paper you don't need. Complain and return to the sender when you get too many layers of packaging.

Look for reusable alternatives to paper products, like organic cotton napkins instead of disposable tissues, reusable cloth bags instead of paper bags, and ceramic mugs instead of paper cups.

You can put paper products in your household recycling bins in most local council areas.

Reduce paper waste by cancelling unwanted advert deliveries, free newspapers and local magazines, and read news online as opposed to buying newspapers. Put a 'no junk mail please' sign on your letterbox to reduce the number of unwanted leaflets and adverts.

Set your printer to print on both sides of the paper. Have a paper policy at your workplace or school which focuses on reduction, reuse, and recycling.

SEE ALSO: Climate Emergency, p. 64; Disposable Products, p. 105; Energy Efficiency and Energy Conservation, p. 116; Packaging, p. 240; Trees and Tree Planting, p. 300

PESTICIDES, HERBICIDES AND AGRI-CHEMICALS

The agricultural system currently relies heavily on chemical pesticides to kill insects and other pests that may eat the crop or undermine the economic viability of the food.

Pesticides are designed to kill. They don't just kill spiders, worms, slugs, flies or rodents; they can also kill any species that eats the insects or rodents. For example, small mammals like field mice are directly affected; birds are also affected because they eat the mice. Tiny amounts of the chemicals accumulate in their bodies and move up the food chain. Herbicides kill unwanted plants by crop sprays targeting certain species that compete with the crop. Herbicides are used to kill plants the farmers don't want among their crop; but they will also kill or damage other plants in the vicinity and, as with pesticides, these chemicals will enter the food chain when animals and insects eat the seeds or leaves that have come into contact with them.

The use of this collection of chemicals is the cornerstone of modern farming. Insecticides are designed to kill insects. Herbicides have the active ingredient to kill plants. Rodenticides are spread to kill both rats and mice to stop them eating crops. Bactericides are chemicals sprayed to kill bacteria. Fungicides kill fungi and larvicides attack the larvae before they pupate.

The residue from spraying crops with these chemical killers also enters the food chain – not just from the food, but because the chemicals leach into the water. This leaching effect means that groundwater, streams and rivers, fish and even drinking water are affected.

Many of the companies that produce the pesticides and toxic chemicals claim that the world would not be able to feed itself unless its products were used to kill the pests. This has been a source of contention and disagreement for a long time. Advocates of organic farming methods admit that the yields may be lower on average but argue that the superior taste and environmental benefits more than make up for it. In fact, even the UN have reported that the world's food distribution and farming problems could be solved with millions more small-scale organic farms across the globe. The problem is that the agricultural industry is now dependent on using these harmful chemicals.

Many pesticide residues can be removed by simply carefully washing your fruit and vegetables. That doesn't make them organic; it just makes the food a little safer for you to eat. The problem is that the farm and farmland where the food was grown has already been damaged.

If we don't eat organic food we are directly contributing to the mass killing of billions of insects and small mammals. The biodiversity and integrity of the farmland is affected and long-term issues arise as the land becomes increasingly sterile.

What can you do?

Avoid food that uses pesticides. Eat organic, buy organic. Don't be part of the problem. Eat a seasonal, plant-based diet that is locally grown.

Don't feed food treated with pesticides to young children and babies who will have a sensitive intolerance to many chemicals.

Support environmental groups who campaign for clear labelling, such as Friends of the Earth; the Soil Association's Food for Life programme, Feed Your Happy, Sustain (an advocacy group promoting healthy practices in agriculture), the Sustainable Food Trust and The Food Commission.

SEE ALSO: Agriculture, p. 27; Chemicals, p. 54; Climate Emergency, p. 64; Fertilisers, p. 147; Grow Your Own Food, p. 177; Habitats and Biodiversity, p. 180; Hormones, p. 196; Packaging, p. 240; Trees and Tree Planting, p. 300

PETS

We love our pets, especially in Britain; we hold them in high esteem and we give them everything they desire.

PDSA, the animal charity, say 26 per cent of British households have a dog – that's about nine million animals. There is no doubt about it: we are a nation of pet lovers, and while dogs may be Britain's favourite pet, we also love cats, rabbits, reptiles, rodents, birds and horses.

In the United States and Canada environmentalists have estimated that pet dogs and cats eat more meat than the entire human population of South America. A huge study by Gregory S. Okin, 'Environmental impacts of food consumption by dogs and cats', set out the grim statistics exposing the hidden cost of pet-food production. The US has the largest population of pet dogs and cats globally, with an estimated 77.8 million dogs and 85.6 million cats to feed.

In China, the number of pets jumped by 8.4 per cent to nearly 100 million in 2019 compared to the previous year, a study on China's pet industry has found.

The impact of meat production on the planet is huge, whether that meat is produced for human or animal consumption. See 'Agriculture' on page 27 and 'Factory Intensive Farming' on page 137.

The vast dog population of Britain also causes other environmental and urban problems. Britain's canines produce 900 tonnes of dog poo every day! It is illegal for humans not to pick up their dog poo in public and you can see people carrying small bags when they take their dog for a walk.

The *Toxocara canis* and *Neospora caninum* parasite lives inside a dog's digestive system and the worms produce eggs. The eggs of the parasite are released when the dog defecates and remain active in the soil for many years, long after the dog itself may be dead and the faeces would have been washed away by the rain. If a small child touches soil that has contained or does contain active eggs they can become very sick. The symptoms of toxocariasis include stomach upsets, eye problems, sore throat and breathing difficulties. Untreated, this disease can cause permanent blindness. This is an expensive problem for farmers as some farm animals have abortions brought on by the toxicity, some get tapeworms and others need to be put down when they come into contact with dog faeces.

The only way to avoid the risk is to pick up dog faeces immediately and dispose of it carefully. Across the UK a dog owner can be fined up to £1000 for allowing their pet dog to defecate without picking up and disposing of the leftovers.

Some animals that are bought as pets, for example parrots and other exotic birds, are actually endangered species. Capturing them in the wild and bringing them to Britain diminishes their already vulnerable populations. Wild animals do not generally make good pets. Why should you have a bird in a cage? Birds need to fly in the open air to be free.

It is estimated by the government that there are around five thousand primates such as monkeys in the UK, but it is illegal under the Dangerous Wild Animals Act 1976 to keep them without a special licence. It is estimated that only 5 per cent of 'dangerous' animals are registered.

Exotic and rare pets, including amphibians, reptiles and invertebrates, are sometimes sold in illegal markets and under the radar of the authorities.

What can you do?

If you are thinking of buying a pet, think hard about why you want it. Can your needs be met by other means? Always buy from a reputable registered dealer.

Don't buy endangered species caught in the wild, like parrots and exotic birds, snakes and other rare animals.

Don't be cruel. Animals and birds that are wild should be left in the wild.

If you have to buy a dog for work or companionship go first to the dogs' home. Unwanted pets have to be destroyed every day in Britain at the rate of 20,000 dogs a year according to the RSPCA.

If you have a pet read tips on how to care for it from the RSPCA, and if you have a dog, make sure to train it properly.

If you have a pet, especially a dog, always pick up their poo from the streets. Use a compostable bag.

Try your dog or cat on vegetarian food. Cats must eat some meat to survive. Ask the Vegetarian Society about healthy diets for your animals. Cats and dogs should really be eating raw food; most eat processed food.

Support animal charities that help dogs and their owners to be responsible, for example the RSPCA or the Dogs Trust, the largest UK charity for dogs.

SEE ALSO: Environmental Stewardship, p. 133; Gaia, p. 165; Zoos and Captive Animals, p. 328

PHONES, COMPUTERS, TABLETS AND TVS

Phones, computers, tablets and TVs are all part of our lives. They are constantly changing, updating and improving. It is amazing what we can do with them today. We can connect all our devices together. Everything is literally a few clicks away.

Research has changed within a generation. Our communications, calculations and visual input have changed dramatically and a vast amount of information is available immediately. Data collection and information flow now power the global economy. Computers drive business, medicine, travel, education, science and almost every aspect of our lives.

However, with all our exciting developing technologies there are environmental issues, health risks and costs.

A single mobile phone can contain up to sixty different materials; including copper, gold, silver, lithium (used in batteries), and indium tin oxide (used as a semiconductor material in device screens). These elements and alloys, known as critical raw materials – are sourced from beneath the earth's surface – they are either precious metals or rare-earth elements. The high demand for and extraction of these minerals have both environmental and social effects on the wildlife and the communities of the areas in which they are mined.

The mining takes place predominantly in China and Africa: China is the lead global extractor of the critical materials for this industry. Mining in Africa is on a smaller scale. However, the day to day management of infrastructure and the welfare of the workers in Africa are very poorly regulated. Greenpeace highlights the hazardous chemicals, dangerous mining practices and badly designed components in this industry, and that many electronic goods now carry a growing number of these minerals.

Mica, a silicate insulator used in a wide range of consumer goods such as TV and radio sets, comes from India. Mining for mica caused deforestation and loss of wildlife and was banned as an industrial process in 2000. Now the mining is done without permission, and by hand, and illegal deforestation continues. Thousands of women and children are exploited in the illegal hunt for mica, which is also used in the cosmetics industry. They are paid just 100 rupees (£1 sterling) for 20 kilos of the raw material. Mica mining causes many health problems, including tuberculosis and other chronic illnesses, for those involved. The Responsible Mica Initiative, started in Paris in 2017, pledges to eradicate mica mining and use by 2022.

The electronics industry is large and voluntary standards are regularly used. Europe-wide standards have helped regulate and process many of the various issues for goods sold within the EU, but elsewhere environmental standards, human safety and wildlife preservation can be poor. Greenpeace International's report 'A Greener Guide to Electronics in 2017' is the most recent edition of their analysis of the environmental impact of the industry's leading brands. The Greenpeace campaign Rethink-IT challenges the IT sector to take full responsibility for its environmental footprint.

We use electricity to power our gadgets and much of our electricity is produced by burning fossil fuels, which generates carbon dioxide and other greenhouse gases.

The US Consumer Electronics Association says the average user lifetime of a mobile phone is between 22 months and 4.7 years. We want to upgrade our phones as soon as we can, but in any case, the phone may have other problems. Obsolescence is built into the design so, for instance, a key function stops working and cannot be repaired so you have to upgrade and sign a contract for a new device.

All electronic devices are harmful if they are not disposed of properly and separated from the general household waste. There are now a number of companies who will buy your electronics from you to deconstruct them for rare metals. These useful metals can be worth more than the phone itself.

As well as the device itself there are all the accessories, wires and cables that went with it, which are reusable – though maybe no longer of use to you – and should be taken to your nearest recycling or repair-and-reuse centre.

Growing health concerns around the subtle effects of non-ionising, low-level electromagnetic radio waves, which are given off by electronic devices, continue and require more research.

Our eyes can be affected if we sit too close to screens for too long. A number of other painful and niggling illnesses have also been associated with the electronic devices we use on a daily basis, such as eye strain, headaches and blurred vision.

Repetitive typing on our devices can also lead to strains to the wrists and fingers and even to diseases such as tenosynovitis and tendonitis, where the tendons of the arm become painfully inflamed. This can be a lifelong problem if you continue to use the computer after you have inflamed fingers, tendons and elbows. Take a break.

There is also growing concern over the long-term effect of the blue light that is emitted from the screens. Such light from electronic sources can cause eye strains, and excessive exposure can cause retina damage according to research conducted by the Prevent Blindness charity. It is understood that children's eyes absorb more of this blue light than those of adults. Trade unions have been campaigning to get safer conditions for people using computers and phones all day at work.

Ultimately we have to merge our computers and screens into one low energy device and at the same time campaign for the greenest and least damaging zero carbon version of these machines in the future. All the old devices that we have been using for the past fifty years should be collected in a national effort so the rare metals can be extracted and used again.

What can you do?

Only buy what you need; reduce your overall need for many devices and machines. Look for environmental standards and company policies that take these issues seriously.

Be cautious and sensible when using electronic devices. Take frequent breaks. Walk away from your screen.

Employees need health and safety tests for using computers at work to make sure they are using their devices in the safest way for their posture, eyes and body. Some employers offer free eye tests if you are using a computer all day.

Make sure when you are working on a computer or laptop that your back is fully supported. The back of the chair should support your whole trunk.

Turn electronic items off when you're not using them.

Watch TV with your family and friends. This saves energy with the double benefit that you are spending time with loved ones.

Recycle all electronics. Some shops accept electronics for recycling, and your local council waste site should take all electronic items. Recyclenow.com is a great UK-based website that will help you to find recycling points, although these will not pay you for your old devices.

Get into the habit of switching your devices off properly.

Support organisations fighting for better conditions for workers and good environmental policies and standards, such as Greenpeace and the Responsible Mica Initiative (RMI, www.responsible-mica-initiative.com).

Use Ecosia, the search engine that plants trees.

SEE ALSO: Climate Emergency, p. 64; Hazardous Waste, p. 186; Heavy Metals, p. 188; One Planet Living, p. 234; Your Carbon Footprint, p. 322

PLASTICS

Plastic is all around us.

Plastic in its many forms has become one of the world's most common substances. A cheap by-product of the petroleum industry, it has replaced paper, wood, metal and many other types of material. Single-use disposable plastics contribute to the climate emergency at every stage of their production.

The Centre for International Environmental Law says that by 2050 single use plastics will be responsible for 13 per cent of the carbon emissions on the planet, equivalent to the pollution from 615 coal fired power stations. It is difficult to recycle, and breaks down in the environment into even smaller pieces which can then be ingested by wildlife and ocean creatures.

The production of plastic really started in the 1950s, and since then a huge 9.2 billion tonnes has been produced, according to a report in 2017 by *National Geographic*. Furthermore, scientists have reportedly found that, of this figure, a staggering 6.9 billion tonnes has become waste, and 6.3 billion tonnes of that never made it to be recycled. Instead, this plastic waste has either been sent to landfill or left in the natural environment, where discarded plastic and other litter kill land animals and marine life every year.

In our rivers, lakes, seas and oceans plastic debris kills around one million creatures each year, a report by the Sea Turtle Conservancy found. Furthermore, eight million tonnes of the world's plastic, according to a study conducted by a group of scientists at UC Santa Barbara's National Center for Ecological Analysis and Synthesis (NCEAS), ends up in our oceans each year, killing around one million sea creatures also.

In the North Pacific Ocean there sits what is known as 'The Great Pacific Garbage Patch' – a collection of 80,000 metric tonnes of waste floating in the ocean. Sir David Attenborough showed us that the plastic contents of this garbage claims the lives of hundreds of thousands of birds and other sea creatures every year on what are supposed to be isolated, human-free Pacific islands. These animals think it is food, as red-plastic bottle tops even mimic the same shape and colour of their usual diet. Until now everything in the ocean was organic matter and could be trusted. Now there is inedible, dangerous plastic, and birds have been feeding it to their chicks.

In order to combat this garbage patch, an Ocean Cleanup device was designed and created by Dutch scientists. This system has a U-shaped barrier with a net-like skirt that hangs below the surface of the water. It moves with the current and collects plastics as they float by. Fish and other sea animals will be able to swim beneath it. The device works by capturing and holding debris ranging in size from huge, abandoned fishing gear, known as 'ghost nets', to tiny microplastics as small as 1 millimetre. In 2019 the non-profit Ocean Cleanup had begun to capture and remove plastic from the Great Pacific Garbage Patch.

The problem with plastic is that it is not biodegradable. In fact, when plastic breaks down, it merely degrades into smaller and smaller pieces, known as microplastics. WWF has documented that it can take 10–20 years for a flimsy plastic bag to decompose in landfill, 450 years for a plastic drinks bottle and at least 500 years for hard plastics, including laminated signs and moulded goods such as toothbrushes.

Some companies claim that they have made bioplastics that are biodegradable or photo-degradable (which break down in light), but the British Plastics Federation assert that none of these substances is truly biodegradable and that the plastic only breaks down into pieces small enough to be invisible to the human eye.

In an effort to combat the plastic pollution caused by single-use items, in 2018 members of the EU voted to ban them. The ban is due to be in place by 2021 and includes items such as plates, cutlery, straws, balloon sticks and cotton buds. The ban will also include products made of 'oxo-degradable' plastics, such as bags or packaging that break down in water and oxygenated conditions, and fast-food containers made of expanded polystyrene.

Despite a 'War on Plastic', plastic use in Western Europe is growing every year and in the UK around 5 million metric tonnes of plastic waste is generated annually according to Our World in Data, yet only 45 per cent is recycled.

Plastic waste can be incinerated in mixed waste to create energy. However, waste incineration sites are invariably controversial for local people as many do not want them near where they live. Plastic and mixed-waste incineration has the potential to emit low levels of toxic pollutants such as dioxins, acid gases and heavy metals, unless safety is strictly enforced.

Throwaway plastic cling film used for food wrapping may contain plasticisers or vinyl chloride. These are poisonous chemicals that can get into cling film-wrapped food, especially if it is warm or fatty. Hard plastics, like kitchen bowls and dustbins, can contain cadmium, which is a heavy metal. Cadmium and its compounds are highly toxic and exposure to this metal is known to cause cancer and targets the body's cardiovascular, renal, gastrointestinal, neurological, reproductive and respiratory systems.

BPA, an industrial chemical found in polycarbonate plastics, is often used for containers, such as water bottles, that store food and drinks. BPA can leach into food and water supplies, and humans are widely exposed to it. BPA is an endocrine disruptor; it can change the body's hormones (see 'Hormones' on page 196).

In 2013, scientists from Brigham and Women's Hospital in the USA published findings showing that BPA exposure can affect egg maturation in humans. Research has also linked even low-dose BPA exposure to cardiovascular problems, including coronary-artery heart disease, angina, heart attacks, hypertension and peripheral artery disease, as found in a report by the US National Library of Medicine National Institutes of Health. Scientists believe BPA, with its oestrogen-like behaviour, could increase the risk of breast, prostate and other cancers in people who were exposed to it in the womb.

What can you do?

We are no longer under any illusion about the problems that plastics have caused in the environment. The product needs to be handled with the same mindfulness and conscious standards that we might use with, say, the disposal of asbestos or heavy metals or hazardous waste.

Reduce plastic use whenever you can, and refuse single-use plastics.

Opt for stainless-steel containers, silicone lids or wax wraps and brown paper bags which are all good reusable alternatives to wrapping food in cling film. Avoid buying new plastic containers, and especially those containing BPA. Always check for a BPA-free label on foods and packaging.

Try to buy and store foods in glass jars, not plastic, and reuse them for food storage.

Repurpose and reuse plastic tubs and pots for non-food-related uses such as flower pots, seed-sowing or storage.

Don't accept plastic carrier bags. Take your own reusable bags or a basket with you when you go out.

Say 'no' to over-wrapped foods.

Don't buy brightly coloured plastics unless you can be sure they don't contain cadmium, a chemical pigment used to give good colouring and high opacity and tinting strength in plastics.

Pick up rubbish when you are out and about; you might strike up a conversation and encourage other people to do the same. When going on walking trips take a clothes peg and a few bags with you for a litter-pick along the roads, verges, beaches and footpaths where you live.

Support The Ocean Cleanup project and City to Sea.

Can you find alternatives for the plastic items you regularly use? Replace your plastic toothbrush with a bamboo one. Order milk from a morning delivery service in returnable glass bottles.

Are you using reusable, washable or sustainable plastic-free baby nappies? Are your children's toys safe to use and plastic-free? Are you still using disposable wet wipes, which are full of chemicals and tiny particles of plastic? Could you switch to washable cloths?

Can you buy your cosmetics – shampoo, body wash and make-up – in refillable options or in packaging without plastic?

Stop using that washing-up sponge – it is completely synthetic and as it starts to wear away it will separate into microplastics. Opt instead for a natural washing sponge made from the loofah plant or a coconut scrubber. Visit your local refill store for natural washing-up liquid available in refill form so you don't have to buy a new plastic bottle over and over.

In the UK, RecycleNow is a fantastic website to use for a quick search for information on how all types of items can be recycled. It is simple to use and based on your location.

TerraCycle offers free recycling programmes funded by brands, manufacturers and retailers around the world to help you collect and recycle your hard-to-recycle waste. Head over to their website to find any waste recycling centres near you.

SEE ALSO: Chemicals, p. 54; Climate Emergency, p. 64; Disposable Products, p. 105; Litter, Waste and Landfill, p. 218; Seas, Oceans and Rivers, p. 278

PLUTONIUM

The most dangerous human-made element ever created is plutonium. Only a tiny amount of plutonium – the size of an orange or apple – is needed to make a nuclear bomb. Worse still, a minuscule amount, too small for the human eye to see, could cause cancer and death.

When a slow-moving neutron is absorbed into a large nucleus (for example Uranium-235), nuclear fission occurs, creating two smaller nuclei (for example Tellurium-135 and Zirconium-97), more neutrons and an enormous amount of released energy.

These neutrons can be absorbed by Uranium-238, a common non-fissile isotope of uranium also found in nature, and can create Plutonium-239. Plutonium can be extracted and separated from nuclear waste in a reprocessing plant.

Plutonium emits alpha particles which have a short half-life but are a highly ionising form of radiation. While only a sheet of paper is needed to stop the radiation, it is still one of the most dangerous elements known to humankind. Radiation from the alpha particles if inhaled or ingested can damage DNA, increasing the rate of cancer.

Plutonium has an overall half-life of 24,000 years; that means it would take that long for half of it to decompose and reduce its radioactive danger. In other words, plutonium will last on the planet for a hundred thousand years.

Countries with nuclear fuel reprocessing plants such as Britain, France and the United States have all built up large stocks of plutonium. It is currently estimated by the World Nuclear Association that there are more than 260 tonnes of plutonium in the world. That is enough to make 85,000 bombs of similar explosive force to those dropped on Hiroshima and Nagasaki during the Second World War.

It is extremely rare to find plutonium in nature but it has been discovered in tiny trace amounts in uranium ores.

Britain has the largest stockpile of plutonium in the world and the Parliamentary Office of Technology and Science estimates it will cost around £73 million per year for the next 110 years to keep it safe.

What can you do?

The Campaign for Nuclear Disarmament (CND) is an organisation whose priority is stopping the world production and use of nuclear weapons.

Plutonium is stored and stockpiled around Britain, most of it in Cumbria under armed guard and above ground managed by the UK's Nuclear Decommissioning Authority.

SEE ALSO: Climate Emergency, p. 64; Fossil Fuels, p. 160; Hazardous Waste, p. 186; Nuclear Power and Nuclear Weapons, p. 229; Radiation, p. 263

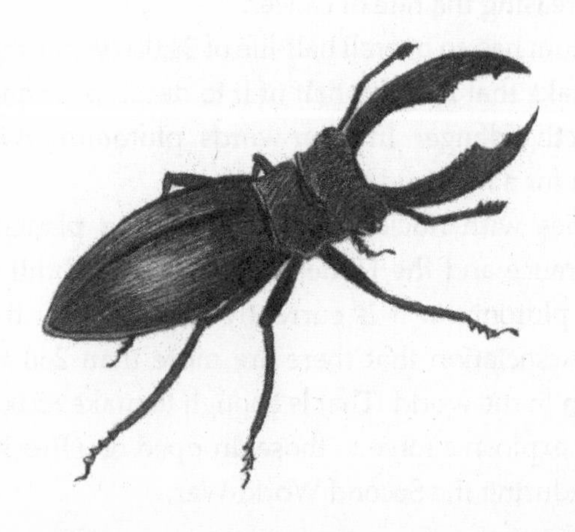

Q

QUESTIONS?

It makes sense to question things. *Homo sapiens*, modern humans, have so far created a world where they are in charge. They have also chosen not to take full responsibility for their power. By choosing to kill and use other species on a global scale we have caused uncountable problems. Species extinction is real.

Humans have hurtled towards our environmental destruction mainly by not questioning enough. It is our right to know the answers and to demand solutions.

Many countries have a law called the Freedom of Information Act. This allows you as an individual to ask questions of any institution, or government body, or public company. Unless the information is an actual state secret, they have to write to you and answer you truthfully as long as your letter contains the phrase 'Under the Freedom of Information Act (FoIA) I would like to ask you about . . .' (or some similar wording), so the request is passed to the nominated FoIA officer to be answered. The law states that the question shall be answered unless the collection of data required would cost too much to prepare. Many journalists and campaigners use the FoIA to have government reveal the truth about issues they care about that have previously been hidden from public scrutiny.

What can you do?

Question your elected MP. What are they doing about species extinction or to protect local wildlife sites? What about your local council, your school, workplace, university? Question them too.

Question those companies that are not on a path of sustainability or fair trade. How can they become more aware? What should they be doing to satisfy you that they are really green?

Allow yourself to have an inquisitive mind. Don't stop yourself or cut off your enquiry. Why should you stop wanting to know more? Don't let others stop you either. There are sometimes powerful forces that don't want a person to be informed.

If you read an advert that says a perfume makes you irresistible, ask if the advert is trying to make you discontented with yourself? Advertising is especially prone to exaggeration and to pinpointing people's insecurities. Be smart, be aware and be inquisitive.

SEE ALSO: Action, Protest and Empowerment, p. 16; Conscious Living, p. 74; Ethical Living, p. 135

R

RADIATION

Radiation is a form of energy transmitted as waves or streams of particles. There are two types of radiation: non-ionising radiation, present in light waves, radio waves and microwaves; and ionising radiation, such as X-rays, neutrons and alpha and beta particles. When radiation from nuclear power stations is referred to, we mean ionising radiation which has a short wavelength and more energy. There are also two kinds of exposure to radioactive substances depending on the wavelength. Non-ionising radiation has a longer wavelength and lower energy, for example ultraviolet light and radio waves, and it cannot get inside the body but can sometimes damage the skin and eyes by burning.

'Background radiation' comes from natural sources, mostly from radon gas, which is a product of natural uranium decay. By far our greatest exposure – 87 per cent – is to natural radiation. The granite of Cornwall emits radon gas, but the risk from this can be very much reduced by good ventilation.

Ionising radiation, a human-made source, is a health risk because exposure to it can cause chemical changes of our body tissues. The changes can manifest themselves as cancer, where a cell mutates, or as disease and damage in future generations because the body's genes have been affected. Radiation increases the chances of cancer developing later in life, and exposure to it is cumulative – every exposure adds to the dose of radiation already in the body. A lot of exposure to radiation can cause 'radiation sickness'; this is what happened to the fire-fighters at Chernobyl who were exposed to a very high dose in a short time.

Children are particularly vulnerable, and cases of childhood leukaemia have been related to radiation from nuclear power stations and plants in Germany. UK and French statistics do not show the same levels of illness and there is more long-term research needed.

The risks of exposure to radiation are hotly disputed. Environmentalists claim that any level of exposure is dangerous, increasing our chances of radiation-related illness later in life. They are particularly concerned that the public may be exposed to radiation from radioactive-waste sites. According to the World Nuclear Association, at the end of 2019 the UK had fifteen reactors generating around 21 per cent of our electricity. The *Guardian* reported that by 2025 the UK aims to shut down the remaining nuclear sites (see 'Nuclear Power and Nuclear Weapons' on page 229). Irradiation is a food-preserving technique using ionising energy at low doses in order to lengthen the time the food is on the shelves in supermarkets (see 'Irradiation' on page 206).

The major earthquake and tsunami in Japan in 2011, which caused three cores from the Fukushima Daiichi Nuclear Power station to melt, caused the world's worst nuclear incident. Fifty thousand households were immediately evacuated and radioactive pollution spewed into the air, the water and the soil. There have already been at least 2,129 radiation deaths from the accident and thousands more are expected over the coming decades. The site itself, and the entire region, is in a state of emergency with radioactive water still leaking into the surrounding oceans.

Fish and vegetables have been at risk and many are inedible. Restaurants have Geiger counters to test food for radioactivity. Life is not normal and it may take many years yet for a return to any sense of normality, if ever. This is the truth of nuclear accidents and their long-term effects.

From X-rays to stone buildings, radiotherapy to nuclear medicine, radiation is part of our lives. Friends of the Earth have made recommendations for reducing the dose limit deemed acceptable for radiation workers and the public. They also argue that the government should provide funds for corrective building and information for houses that have a high level of naturally occurring radon.

What can you do?

Radiation is a complicated issue. To find out more, contact Friends of the Earth, Greenpeace or CND.

If you don't like nuclear energy then support renewables which are cheaper and safer.

If you live near a nuclear power station, reprocessing station or waste site, you can join a local environmental group and work with them to campaign for greater safety.

SEE ALSO: Nuclear Power and Nuclear Weapons, p. 229; Plutonium, p. 259; Renewable Energy, p. 270

RECYCLING

Recycling is the process of putting waste materials back into the system to be transformed into a new product. Packaging like glass bottles, cans, paper, metals and many textiles can be recycled in an industrial process and used as raw materials for the manufacture of new goods and products.

Recycling leads to a reduction in the environmental damage caused by the production of goods. It also means there is ultimately less mixed waste in landfills. It can lead to job creation in collection schemes, recycling workshops and stores of material to be sold for reuse.

Recycling makes economic sense.

Glass, plastics, newspapers, cardboard packaging, food containers and yoghurt pots can all be recycled. Some products, like glass, can be endlessly melted down and reused, saving energy and raw materials. It is well known that the recycling of one glass bottle can save enough energy to power a computer for 25 minutes.

New schemes are opening in Scotland and in smaller county regions to encourage reuse and recycling by including in the price of bottled drinks a deposit you reclaim when you return the empty bottle to the shop. According to the Campaign to Protect Rural England, 72 per cent of the public would support the reintroduction of a UK-wide deposit return programme for bottles. This is one practical way the consumer to producer cycle can be closed.

Recycling aluminium makes even more economic and environmental sense. It takes the same amount of energy to recycle twenty cans as it does to mine and manufacture one new one.

In Yorkshire in 2018/19 the East Riding local council achieved a 64.8 per cent score for recycling, reuse and composting rates, the highest rates reached in the UK. Overall the UK has a 45 per cent average. Germany has the highest recycling rate in the world with 66 per cent of their entire waste stream being saved from landfill having met their own Federal Government target of 65 per cent target in early 2020. The

Welsh government has the fourth highest recycling rate in the world with 52.2 per cent.

Some new council waste systems are able to recycle more than 95 per cent of the waste stream if the public would fully get behind them. Waste and recycling enthusiasts have proved that a family can live well and produce only a tiny jar of non-recyclable rubbish in a month. It is all about avoiding unnecessary household purchases, increasing redistribution and recycling, and household members making a conscious and concerted effort together to take waste to other places to be reused.

Natural resources all over the world are under pressure. Our industries and the waste we produce are causing terrible pollution. Recycling is a logical way to reduce the damage to our planet and it has the added bonus of saving us money.

Once collected and sorted, recyclable materials become valuable commodities in the worldwide market.

The biggest problems occur when items are put in the wrong recycling container or when containers are contaminated with food. Households have to be aware of their own part in cleaning their jars and cans and making sure not to cross-contaminate a batch with the wrong item.

There are now a number of high-street retailers who offer textile recycling schemes, and some even give the donor an incentive in the form of a gift voucher or discount code.

Some retailers also encourage customers to bring back containers to stores for proper recycling – usually hard-to-recycle items such as squeezy tubes and multi-material containers.

TerraCycle offers free recycling programmes funded by brands, manufacturers and retailers around the world to help you collect and recycle your hard-to-recycle waste. Head over to their website to find any waste streams near you.

We must start recycling now!

What can you do?

There's a lot you can do in this area. Be inventive. How many things do you throw away that could be reused or repurposed into something new? Upcycle and redistribute items as the first alternative to recycling.

Avoid disposable and single-use products where possible.

Consider buying second-hand items from charity shops, online websites or even swap between family and friends, rather than buying new things. Similarly, give unwanted things or outgrown clothes to a charity shop or to a high-street retailer's recycling bank.

Get familiar with and understand recycling symbols. Which materials are easier to recycle and are accepted by your local council? If anything is not accepted within your waste centre, don't just throw it in the general waste, share it, give it away, or find a local reuse or recycling point.

In the UK, RecycleNow is a fantastic website to use for a quick search for information on how all types of items can be recycled. It is simple to use and based on your location.

Investigate TerraCycle and start collecting recyclables for one of their programmes; the more you collect the more money can be raised for charity.

SEE ALSO: Consumption and Consumerism, p. 77; Litter, Waste and Landfill, p. 218; Your Carbon Footprint, p. 322

REFUGEES AND HUMAN DISPLACEMENT

Because of political turmoil, persecution and violence, millions of people have to leave their homes and find safety elsewhere. They become refugees. Sometimes they have to run away, often without even a bag or personal items. Most move close to their original home. About 50 per cent of refugees are under the age of eighteen according to the United Nations Refugee Agency, UNHCR.

UNHCR reported that 70.8 million people were persecuted in countries through violence, conflict and human rights violations during 2018 and more than 3.5 million asylum seekers around the world moved from their own country. They were mainly children and families fleeing from danger.

Britain takes in the lowest number of refugees of any European country; only 0.25 per cent of our population are refugees. In 2018/2019 Britain received about 35,000 applications, but the process is rigorous, the rules are tough and even children who travel on their own are not automatically accepted.

Life is hard for a refugee. Many are traumatised by what they have seen or experienced. Many are held in immigration detention centres until their cases are heard. At any one time the Refugee Council is working with more than 20,000 people who are waiting for longer than six months to have their legal cases heard in the UK. They are not allowed to work or earn money during that time and therefore suffer severe financial hardship.

The climate emergency has created even more refugees and the numbers are increasing. When drought, fires and severe heat destroy a country people move or they will die. When there is no water, life itself is at risk.

It is difficult to say how many climate refugees there are. Climate change has a complicated impact on drought, on farming and growing food. Some people leave their homes because of lack of food or water. Some leave immediately when there is a fire or flood.

The situation is going to get even more difficult for years to come, especially if we don't do anything to help. The United Nations say millions of people will be displaced in the future because of the climate emergency.

What can you do?

Do you have compassion for refugees? How can you be supportive and not judgemental? How can you bring love into the situation? You could support one of the following organisations:

- The Refugee Council – supporting refugees and asylum seekers in Britain
- The Climate and Migration Coalition – working with people who are displaced by climate change
- The Children's Society – supporting unaccompanied child refugees
- UNHCR – the UN Refugee Agency – offering legal advice to refugees in the UK
- Women for Refugee Women – supporting and empowering refugee women

SEE ALSO: Climate Emergency, p. 64; The Development of Humanity, p. 102; Ethical Living, p. 135

RENEWABLE ENERGY

Renewable energy is one of the fastest growing businesses in the world today. The renewable energy sector bypasses the carbon problem of fossil fuel burning that causes 50 per cent of the carbon dioxide in the atmosphere. Renewable energy sources produce clean energy. There is almost no pollution from them while they are generating energy, and they can be repaired and upgraded to ensure a long life.

Renewable energy encompasses wind power, solar power and power from water turbines, among other sources of efficient and nature-driven energy. Globally, one third of our power capacity is already provided from renewable energy sources and this is growing.

Coupled with energy efficiency measures, the renewable energy industry has the capability to supply all the energy Britain needs. We need a political will to bring that about.

Other countries have already done it. Iceland now provides 100 per cent renewable energy from nature: 87 per cent of their energy comes from hydro-power and 13 per cent from geothermal power. Geothermal power comes from stored heat in the rocks deep under the earth.

In Costa Rica, a country of five million people, 99 per cent of their supply is renewable energy, coming from geothermal, hydroelectric and wind power. Wind power is popular in Scotland: on many days in the past few years Scotland has produced 100 per cent of its energy needs from wind.

Love the sun! The sun is, of course, the source of all energy originally. *Forbes* magazine tells us that the price of solar photovoltaic cells is now so low that solar power is cheaper than all the fossil fuel alternatives. These cells are expected to last for twenty to thirty years at least and longer lifetimes are increasingly possible in newer designs. In Africa solar power is changing lives and lifting people out of poverty. Children are able to get their homework done by solar electric light instead of using kerosene lamps, which are unhealthy, expensive and dangerous. The AfDB Solar Project has provided solar power to ninety million people in Africa.

Wind power generates electricity without emitting toxic pollution and carbon dioxide which kills millions of people. The wind will blow for free; it makes sense to harness this energy form for good. It is now an economically viable alternative to fossil fuels. Wind turbine manufacturing accidents and human deaths globally have been calculated at around 49 people per year by the Wind Energy Industry.

Bird deaths figures from wind turbines in the UK have been examined after some world leaders tried to disparage wind energy as a viable alternative to coal. The London School of Economics suggested in 2014 that between 9,600 and 100,000 birds could die from wind collision accidents by 2020. The BBC ran a follow up report in 2019 and concluded: we are not sure.

This figure pales into insignificance when we learn that some 55 million birds die each year in the UK because of household cats. Cats are the number one killer of birds globally; a study by British scientists published in the *Journal of Applied Ecology* in 2013 estimates that cats account for 2.4 billion bird deaths every year. Buildings and glass windows are responsible for a further 365 million to 1 billion dead birds in the US according to Smithsonian researchers, and cars follow with nearly 200 million deaths.

Renewable energy is the future.

What can you do?

If you can, link your household with an electricity supplier that uses renewable energy such as Green Energy UK.

Save energy when you can and adopt energy efficiency measures in your home: the more you save the less pressure there will be to keep finding new forms of energy.

Visit the Centre for Alternative Technology in Wales. This pioneering community is powered 100 per cent by the sun and wind.

SEE ALSO: Climate Emergency, p. 64; Energy Efficiency and Energy Conservation, p. 116

S

SANITARY PRODUCTS

There is still some shame and stigma around menstruation. Millions of girls globally struggle with the social, economic and environmental effect of periods; in many instances this is now referred to as period poverty. Every month one in three girls in Bangladesh, Pakistan and Sri Lanka miss school days during their periods according to WaterAid. This is mainly because of a lack of toilets and access to period products like sanitary towels. When girls are bleeding they are expected to stay at home. In remote parts of Nepal, girls and women are sent to a bleeding hut until their period is finished. Some have died due to the harsh conditions.

Periods are natural, yet the lack of basic resources and support means young girls and women are held back from reaching their full potential. They are kept away from school or work for instance when they are on their period. Charities such as ActionAid, PLAN International, WaterAid and freedom-4girls UK are doing some truly inspirational work to help girls around the world overcome this stigma.

In the UK, industry statistics show that 4.3 billion disposable menstrual products are used every year. There are environmental issues regarding disposable menstrual products and there is also concern about the levels of toxic chemicals, chlorine, plastics and micro-absorbers found in sanitary products.

Well-known health effects associated with the use of tampons include toxic shock syndrome (TSS), vaginal ulcerations and dryness. TSS is a rare but serious disease; the symptoms include high temperature, headache, vomiting and diarrhoea, a sunburn-like rash and a rapid drop in blood pressure, which in some cases can lead to death. It is particularly seen in women using tampons. More detailed information on TSS is in the instructions inside the tampon packet.

Product manufacturers say that a woman may use around 11,000–16,000 disposable sanitary products in her menstruating lifetime. It is estimated that around 700,000 panty liners, 2.5 million tampons and 1.4 million pads are used every day in the UK, and, according to figures published by the Institution of Environmental Science, many are being flushed down the toilet, ultimately blocking our sewage systems and even escaping into rivers and seas. Never flush your sanitary products down the toilet.

City to Sea, which campaigns to reduce plastic pollution, found that sanitary products make up an estimated 6.2 per cent of beach litter in the UK and are the fifth most common item found on Europe's beaches.

The Women's Environmental Network (WEN) launched a campaign to publicise the environmental consequences of sanitary towels and tampons in 1989.

In 2017, WEN teamed up with City to Sea to launch a global campaign, 'Plastic-Free Periods'. This campaign works to raise awareness and educate on health-conscious, environmentally-friendly menstrual products, aiming to reduce plastic pollution from period products. By 2018 the campaign had reached two million people and in 2019 City to Sea began a nationwide, unbiased schools programme calling on the Department of Education to only purchase plastic-free period products for schools – to provide girls with products that are better for their bodies, the planet and their future.

It is now possible to buy disposable sanitary products made from natural rather than chemically produced materials. Manufacturers are producing non-chlorine-bleached, organic cotton sanitary products, including towels and tampons, which are wrapped in either compostable paper or plant-based bio-wrappers.

There has also been development on reusable sanitary products. Reusable cotton pads, super-absorbent knickers, and menstrual cups offer alternatives to disposable products for absorbing the bloody liquids of periods. For tampon-users, reusable tampon applicators reduce the amount of waste created. These are all becoming more available, even in high-street supermarkets, and via many companies online.

After thirty years of campaigning by women's organisations, the 'tampon tax', an arbitrary 5 per cent tax on sanitary products, was scrapped in January 2021.

The Government of Scotland has been supplying all schools with free sanitary wear since 2018.

What can you do?

Join the Women's Environmental Network and City to Sea, and support their campaign.

Join or support The Good Goddess Project, which collects money and items of environmentally-friendly sanitary wear for destitute women and refugees across Britain and overseas.

Always use natural and organic products, whether reusables, washables or disposables. Choose products with the least packaging and avoid any plastic; always read the info on the box to see what's inside.

Change menstrual products frequently and do not use a single product for more than eight hours, due to bacteria build-up. Always use the lowest absorbency that you need, i.e. regular rather than super. Use towels or reusables at night rather than tampons.

Consider making your own reusable pads. There are dozens of free videos and templates available online. They can be sewn together in about an hour and, of course, they are free, using scraps of cloth.

Never flush sanitary products down the toilet; always use the bin.

SEE ALSO: Action, Protest and Empowerment, p. 16; Chlorine, p. 58; Consumption and Consumerism, p. 77

SEALS AND RIVER LIFE

When the river is clean the animals and river mammals flourish. In the 1950s the River Thames was so dirty that no seals existed or could survive there; it was considered biologically dead. There was a public outcry and pressure groups and charities campaigned to change the law.

Legislation changed the way we disposed of many pollutants. That has made a huge difference. In fact, in the past few years seals have returned to the River Thames and many of the other rivers in Britain. Clean river water allows the whole river biodiversity to flourish and the larger animals like seals have enough food to eat. They mainly feed on squid, crabs and whelks.

In 2019, 138 harbour seal pups were born in the River Thames for the first time in more than 60 years. This was the first decisive investigation of the creatures that have been tracked and counted by conservation volunteers at the Zoological Society of London. There was much rejoicing. Now it is estimated that rivers around Britain have been cleaned to such an extent that river life is doing very well. There are thought to be more than 3,500 adult seals in the Thames, and thousands of others across the country, in estuaries and harbours and along the coastline.

Seals are an important indicator of a sustainable river; it means that many other species must be thriving. They are joined by otters, water voles and beavers who are returning to our rivers. Kingfishers, delicate river birds and woodpeckers are also returning. The clean water is important to allow birds and fish to thrive and the ecosystem to flourish.

The UK has more than 1,500 river systems, many joining together. The River Severn is the longest at 354km, followed by the Thames, which courses through London and is the country's most famous and deepest river, at 346km long. Rivers are the most important water systems on our land, and we must keep them clean and protect them for the millions of people and species that depend on them.

What can you do?

Support river and wildlife charities like the Wildlife Trusts, especially your local Trust.

Never throw rubbish in any river. Support The Rivers Trust and the River Thames Society.

Organise river litter-picking in your area, especially if you spot any fishing items like nets, which wildlife can get caught up in.

Don't feed wild birds or river animals your old and mouldy bread which swells in their stomachs and is not nutritious for them. Instead bring seeds and worms which is more appropriate food.

If you come across seals and their pups on the beach, particularly on secluded beaches, they are probably digesting their food. Keep dogs, other pets and children away. They do not like to be disturbed and they are easily upset.

Adopt a seal: you can pay to support a charity that will protect your chosen seal and its family, and help pay to continue its conservation work in the seal's natural habitat. The Cornish Seal Sanctuary, part of the Sea Life Trust, and the Scottish Wildlife Trust are just two organisations that manage the adoption of baby seal pups. Funding goes to support wider conservation projects.

SEE ALSO: Habitats and Biodiversity, p. 180; Seas, Oceans and Rivers, p. 278; Species Extinction, p. 285; Water, p. 317

SEAS, OCEANS AND RIVERS

The seas, oceans and rivers make up the majority of water on the planet. They are the home to millions of species of fish, mammals and other aquatic life.

Because of human interaction these waters have many environmental problems that are new. Marine life is under threat as never before, from overfishing to pollution by oil spills, sewage, waste, poisonous algae as well as oxygen pollution. The effects are devastating.

Overfishing has caused the disappearance of many aquatic species. As human populations have risen, with the corresponding rise in demand for food, sea-fishing vessels have become so large they are called floating factories. The techniques used in commercial fishing capture many other sea creatures along with the species we actually eat, and most of those creatures die as a consequence. These fishing vessels have been literally hoovering up all the sea life area by area.

The UN Food and Agriculture Organisation say that if we continue at this pace seafood stocks for human consumption will run out by the year 2048. Marine predators, such as sharks and other large fish, are also disappearing. In fact, in the last fifty years we have managed to wipe out 90 per cent of the oceans' large predators. More than half of all shark species are threatened or near-threatened with extinction. Over 97 million sharks were killed in 2010 with real numbers estimated to be in the region of up to 273 million, though there is insufficient data according to a study by Canada's Dalhousie University.

Fish farms are no better for the environment either. They can be an ecological hazard as they allow polluted water carrying pesticides and fish that have had antibiotic treatments and waterborne diseases to enter the rivers and seas. Some farmed fish have lice and have carried infections to other species. A number of the farms have been responsible for huge oxygen-depleted areas of green slime appearing on the surface of nearby waters, which kill aquatic life below. More importantly, it takes two and a half kilos of wild fish, fish oil and meal, to feed the farmed fish and produce eventually only half a kilo of salmon.

As stated earlier, our rivers, oceans and seas are also full of human-produced waste, mostly plastics, which pollute and cause confusion and consternation for fish and seabirds who regularly mistake it for food. The Ellen MacArthur Foundation is worried: 'The ocean is expected to contain 1 tonne of plastic for every 3 tonnes of fish by 2025, and by 2050, more plastics than fish (by weight).'

Sewage is a major problem for the world's waters also. Sewage pollution from our toilets can be flushed out to sea from waste treatment plants. When people put sanitary towels, condoms, paper, wet wipes and plastic waste into the toilet it can end up in the oceans, along with the sewage, depending on the complexity of the sewage treatment. The WWF reported that more than 40 per cent of British rivers and estuaries are polluted by sewage.

Climate change affects aquatic life in many ways, and we are seeing its effects through the studies of coral reefs. These reefs are the large, hard, exoskeleton structures of invertebrates called coral polyps, which are glued together by calcium carbonate; they are vital to biodiversity and sea life, and are the rainforests of the ocean. They are found mainly in shallow tropical waters. Britain has some coral reefs, protected since 2013, which are cold water reefs existing mostly around the waters of Scotland.

Coral reefs are home to one quarter of all life in the oceans, even though they make up only one per cent of the seabed. According to conservation group Secore International, fifty per cent of the world's coral reefs have died in the last thirty years and at the current rate of pollution it is estimated that 90 per cent will die in the next twenty years from the warming oceans, increasing acidity and water pollution from toxic industries.

While coral reefs have taken more than ten thousand years to grow, humankind has destroyed them in the last fifty years.

Pollution, overfishing, fishing with explosives, coral mining, intensive beachside development and tourism are all exacerbating the problem. In addition, corals are becoming bleached from the toxic chemicals leaking into the waters and other pollutants. Bleaching leaves corals with a white-looking exoskeleton and ultimately leads to their death while the fish inhabiting the reefs are declining in numbers. This toxicity affects all life inhabiting reefs by impairing the growth and photosynthesis of green algae; inducing defects in young mussels; damaging the immune and reproductive systems and causing deformity in young sea urchins; decreasing fish fertility and reproduction and creating female characteristics in male fish.

Tourism is also a factor in the destruction of corals. During reef dives it is explained that visitors should not touch or get too close to the corals, but many do not listen to this instruction. By simply touching coral with your fingers, the oils from your skin can disturb the mucous membranes that protect them from disease, and it can even shock and kill them.

Additionally, the lotions we use to protect our skin from the sun's harmful rays are contributing to coral destruction. Toxic chemical ingredients are used in many SPF lotions, and have been directly linked to reef bleaching, deformity and death, especially in young corals. These chemicals prevent corals from maturing by disrupting their hormone development.

Reef-destroying chemical sunscreen ingredients include: Oxybenzone, Benzophenone-1, Benzophenone-8, OD-PABA, 4-Methylbenzylidene camphor, 3-Benzylidene camphor, nano-Titanium dioxide and nano-Zinc oxide.

National Geographic reports that 14,000 metric tonnes of sunscreen is washed into the oceans every year from swimmers and other beach users. The toxic chemicals leak into the waters when we swim, shower off or use spray sunscreen from an aerosol can or plastic bottle. Even the sand can absorb the sunscreen and eventually it enters the water too. This adds to the breakdown of the oceans' ecosystems.

Fortunately, in recent years we have seen some places banning conventional harmful sunscreens; among them are the US Virgin Islands, Hawaii, Key West in Florida, Bonaire and Palau. This is definitely a step in the right direction.

Finally, it's important to state that when the ecosystem of the sea is clean and pure the oceans provide so much food for us to enjoy. Billions of creatures enjoy the fruits of the sea. Sea vegetables such as kelp, kombu, arame, nori and chlorella are all extremely nutritious. They are rich in minerals that help detoxification of the blood, and are low in carbohydrates and fats but do contain protein for brain food. Spirulina is an easily digested immune booster which is antifungal and antibacterial. Sea vegetables contain essential elements such as potassium, zinc, iodine and selenium and are used as a medicine to absorb heavy metals and toxins, increase the amount of minerals in the body, and alleviate illnesses such as stomach tumours, ulcers and cardiovascular diseases. They are excellent at building cell membranes, regulating the digestive system and cleaning the body of radiation and hormones.

What can you do?

Respect the world's waters; don't litter, don't overfish, don't eat too much fish. If you eat fish, eat it as an exceptional meal, a special treat.

Or stop eating fish altogether or at least until fish stocks return to a sustainable level. Eat sea vegetables instead which are in plentiful supply in clean seas and are freely available around the UK coastline.

Be informed. Visit Overfishing.org for more information and advice. The Marine Conservation Society has produced a guide to purchasing fish: 'Guide to Good Fish'. Greenpeace has produced an International Red List of seafood species.

Don't reef dive, and if you do, don't ever touch them. Support UK reef trusts such as the Cornwall Wildlife Trust and The Wildlife Trusts UK.

Don't buy sun lotions from retailers who knowingly add toxic chemical ingredients to their products which are harmful to both marine life and your own health. Use your buying powers and use reef-safe sun lotions which have a reef-safe symbol on the bottle. There are brands such as Badger Organics, Green People, Organii, Amazinc and Shade.

SEE ALSO: Chemicals, p. 54; Climate Emergency, p. 64; Cosmetics and Toiletries, p. 80; Hormones, p. 196; Seals and River Life, p. 276; Species Extinction, p. 285; Water, p. 317

SOIL

A major study from the Intergovernmental Science Policy Platform on Biodiversity (SPPB) has gathered five hundred experts from fifty countries together for the last ten years. They have shown how tens of thousands of species are in danger principally from the over-exploitation of the land and the few fragile and precious centimetres of soil that live under our feet.

Soil, of course, is a mixture of so many things, principally organic matter and bacteria, both living and dead. Then there are minerals, gases, liquids, stone and crystals all sharing the space of the soil. An uncountable number of the earth's creatures live in the soil: insects and worms, weevils and spiders; so many creatures who work tirelessly to be part of the great churning of the largest linear compost piles.

The soil of the earth is exhausted. The SPPB study shows how countries are pushing nature's ability to grow food at such a speed, that the soil is outpacing its ability to renew and sustain itself. The scientists are sounding the alarm. They see compromised land and soil in every single region on earth.

The report shows that all life systems are provided free by nature: clean water, air, soil to grow crops in and the amazing pollination of those crops by insects and bees which keeps the carbon cycle going.

Soil erosion is now at a critical phase.

When drought strikes the soil is stressed even more. Cutting trees exacerbates soil erosion too.

Other quality issues arise when the soil is stressed, for

example by compaction or loss of structure. Alterations in the soil quality can change the entire local ecosystem.

There are three main tests to assess soil health. These are the tests farmers use.

1. Look at the soil. Make a good assessment of its structure, its feel, and evaluate its overall appearance. What is growing? Thriving?
2. Count the earthworms. A healthy soil is full of them. They should be coming up if you move the soil. There are surface dwellers, mid-depth dwellers and deep-burrowing worms.
3. Assess the drainage. In farming it's called an infiltration test. You are looking at the structure of the soil and making sure it can both drain water at speed and also absorb enough water to feed plant roots.

You can check the pH of the soil (its acidity or alkalinity) with a simple kit.

Essentially good soil is one where there is lots of strong healthy organic matter. You can improve soil with compost, and you can make the compost yourself from your own food waste. Improving the quality of soil is critical and an important way in which you can leave the planet better than you found it.

What can you do?

Touch the earth. Get your hands dirty and discover the soil.

Good soil health is critical for our ecological and sustainable survival. If you haven't already, start a compost heap.

Feed your soil, even in the smallest garden, even in a flat with a few pots, with compost and vegetable waste.

Touching the soil is good for your overall health and for the amount of endorphins in your emotional and central nervous system.

Never drop cigarette butts into the soil, the earth or in flowerpots. Cigarette butts are the most toxic form of litter and dangerous for healthy soil.

For soil erosion and degradation information see www.worldwildlife.org

The Permaculture Association and *Permaculture Magazine* have a huge variety of information, courses and films about soil health and making it better for the planet.

See GREATsoils: the Worm Count Test on YouTube.

Get soil information and practical ideas from the Soil Association and the Sustainable Soil Initiative from foodtank.com.

SEE ALSO: Agriculture, p. 27; Climate Emergency, p. 64; Fertilisers, p. 147; Flowers and Plants, p. 153; Grow Your Own Food, p. 177; Habitats and Biodiversity, p. 180; Pesticides, Herbicides and Agri-Chemicals, p. 244; Species Extinction, p. 285; Your Carbon Footprint, p. 322

SPECIES EXTINCTION

At least 2,000 animal species are becoming extinct every year for certain. This number has increased dramatically. Some scientists say that we have underestimated this count and if there are 100 million species on earth, then around 100,000 species are becoming extinct each year.

The number of animal, bird, fish and insect species that have become extinct due to the encroaching climate problems around the world has been growing exponentially. From deforestation to industrial pollution, so many human actions have affected other living species.

We are killing animals on a massive scale every single day.

Our lifestyle, our use of resources, and creating businesses around the death of animals means that we rely on animal products for our survival just as we did thousands of years ago.

Our society, however, has created ingenious and dastardly ways to murder animals. Humans kill for 'pleasure' when people go hunting for wild game. Emotionally, we kill from a place of fear when we swat a bee. We kill for food when we eat cows stunned in an abattoir. Tourists buy trinkets at the beachfront, little knowing that they are buying an endangered species. From tiny daily actions to manufacturing and government policies, animals of all sizes are under attack.

The numbers of animals, insects and other creatures killed by mass clearing of forests is a crisis. Forests that have existed for thousands of years in South America have been cleared in just one generation to enable soy farming to feed cattle for the growing market for beef burgers. Forests in countries like Canada have been cleared on a vast scale to chip the trees for pulp and paper. This paper is for advertising and flyers that end up on your doormat. You probably only look at them for a second before throwing them away. These kinds of businesses exacerbate the already stressed ecological conditions and make it difficult for local native species to survive. They lead indirectly and directly to more species extinction.

Food production is also responsible for animal extinctions. Animals and insects of all kinds are sprayed with poisons in fields and hedgerows across Britain and around the world. This is considered by business to be normal. Birds eat the dying insects; they absorb the poisons themselves and die. Their offspring are unable to grow; the numbers of insects and birds drop, field by field, valley by valley. This has been known about for fifty years!

The weather and climate emergency is also directly affecting species numbers and stress levels. According to the University of Sydney, 500 million animals died in the Australian bush fires of 2019 and the beginning of 2020. That included 30 per cent of the koala species – some 8,000 koalas – that were burned to death in the fires in New South Wales. The fires that spread across Australia caused extinctions by wiping out some endangered species and further stressing and endangering others. Scientists say the ecological devastation will be impossible to repair, such was the scale of the damage and the loss of wildlife.

This is an emergency.

When species decline and become extinct we will never see them again. This current age is being called by biologists and scientists the Sixth Mass Extinction and is a direct result of human activity. It really is that serious.

There is some good news. When protection measures are in place animals do survive, and thrive. Because of the Endangered Species Act, sea turtles, which were once in peril, have made a staggering recovery. A whopping 980 per cent increase in numbers has been recorded because a programme protecting turtles was identified and successfully implemented according to the US-based Center for Biological Diversity.

The numbers of humpback whales were declining and thought to be only 450 worldwide in the 1960s. International protection programmes that stopped countries like Japan from killing them have now seen a massive 25,000 increase in numbers according to research by the journal *Royal Society Open Science*. It took pressure from environmental groups, international laws and determined lawyers and campaigners to make it happen.

What can you do?

We must wake up to the devastation of this loss. How vain is humanity, believing it can replace the bee? Or the earthworm? Or the blue tit or wren? Or a dolphin?

This loss, this emergency, is a tragedy. This loss is overwhelming to the planet. Why are we not weeping? Why do we not read out their names in public with shame and sorrow? Where are the sanctuaries, the noble marble plinths of remembrance with inadequate words of regret?

Call a truce. Decide not to hurt or murder or kill animals any more. Don't take part. Find love instead. Work to protect, to allow survival. Observe how every day society allows and encourages the killing of animals.

Protect animals.

Buy organic. Don't buy animal-related products if you can.

Join groups that fight for animal rights such as PETA. Join groups that work towards compassion in farming methods.

Read this book. We are all interrelated. What kind of life you lead affects the world. Be part of the solution, not the problem.

Hold a funeral to honour all the species that have become extinct. Use the time to reflect and commit to supporting other living species who can't speak up themselves. What are you going to do?

#SixthMassExtinction

SEE ALSO: Agriculture, p. 27; Animals in the Textile Industry, p. 40; Animal Testing, p. 45; Climate Emergency, p. 64; Grow Your Own Food, p. 177; Habitats and Biodiversity, p. 180; One Planet Living, p. 234; Seas, Oceans and Rivers, p. 278; Your Carbon Footprint, p. 322

SUSTAINABLE TEXTILES

Vegan leather from wine grape waste, performance wear made from coffee grounds, ropes and mats made from banana fibres and knitwear from coconut hair? This is how zero-waste policies are creating sustainable textiles and new markets.

Thrifting, swapping and upcycling is a growing market. Statista reported that in 2018 the global market value of the second-hand clothing market was estimated to be $24 billion, according to research conducted by ThreadUp and retail analytics firm GlobalData. Whether it's buying clothing, homewares or vehicles, shoppers are becoming increasingly savvy. According to *Forbes*, conscious consumers are demanding sustainability, and are putting pressure on big name brands, causing them to search for greener solutions.

Many people no longer want to wear shoes made from slaughtered animals and similarly do not want shoes made from plastic either.

Sustainable textile production is obviously the future of the fashion business. At the moment ethical products are often more expensive; the fast-paced high street is detrimental to environmentally sound 'slow' fashion.

But we are seeing more original textile alternatives as people experiment with biodegradable materials. These are the truly innovative and sustainable textile alternatives being used today, which will shape the future of the textile industry.

The coronavirus pandemic saw retailers needing to adapt to the sudden changes we experienced worldwide. Inventive ways of production saw fashion retailer Kitri introduce a pre-order collection, with the idea that they only produce what there is a demand for, in an effort to reduce waste. In a statement Kitri explained that the collection was 'a way of thinking which will minimise waste, surplus stock and create equilibrium between supply and demand'.

Below are listed some sustainable fabrics, and innovative businesses.

Bamboo

Bamboo fibre is extracted from the extremely fast-growing bamboo plant. It is naturally hypoallergenic, absorbent, quick-drying and antibacterial. Bamboo is widely used: from construction to home wares to textiles.

To keep up with ever growing demand, some farmers have been known to use pesticides; and mass bamboo plantations cause deforestation. Always be sure the bamboo you are purchasing has been grown sustainably.

Econyl

Econyl, a yarn created by Aquafil, is a recycled fabric, made from waste such as plastic, fabric and fishing nets discarded in our seas and oceans, which are recycled into a new nylon yarn.

Recycling and upcycling uses less water and creates a lot less waste in comparison to traditional nylon production.

Care for Econyl is particularly important, as due to its fibre origin, it can still shed microplastics during washing.

Mycelium and Mushrooms

Mushrooms really have magical abilities when it comes to producing a fully sustainable alternative to animal leather. Companies such as Ecovative Design and Bolt Threads are creating new technology to replace plastic and reduce the use of animal skins, by turning mycelium (spores) of fungi and mushroom by-products into vegan leather. The wearable fibre is durable, heat resistant, resilient, breathable and of course biodegradable.

Mycelium needs very little water for growth and provides nutritional soil for other plants.

Organic Hemp

Hemp is one of the world's most versatile natural fibres. It is a fast-growing plant which needs little water and does not require the use of herbicides, pesticides or synthetic fertilisers. Hemp's deep roots prevent soil erosion and the leaves, when shed, act as a natural soil improver. It can be used for fabric, rope and for when very tough materials are required.

It is also durable, resilient and naturally antibacterial. Its antibacterial properties could create opportunities for health-care fabrics.

Organic Linen

Linen, an established natural fibre, is derived from the flax plant, a versatile crop. The crop requires minimal water and pesticides and has the ability to grow in poor quality soil, making it very low maintenance.

In addition, there is no waste when it comes to the flax plant: every part of the plant is used.

Linen fabric is hardwearing, and when it is not treated synthetically it is fully biodegradable.

Pineapple fibres

A tough fabric made from pineapple leaf fibre is becoming popular with footwear manufacturers as a sustainable vegan option.

The fibre can be used to produce a natural and non-woven textile known as Piñatex, originating from the Philippines. It is antibacterial, highly absorbent and has antiseptic properties in addition to being extremely lightweight.

Recycled and Upcycled

Recycled and upcycled fibres are made from industrial and consumer waste, making it the most sustainable alternative to both conventional and organic natural fibres and synthetic fibres. Be inspired by brands such as Turtle Doves and E. L. V. Denim.

Red Wine

Innovative creators Vegea have collaborated with Italian wineries to develop a process to produce vegan leather from grape skins, stalks and seeds which would otherwise be discarded during the production of wine.

Spider Silk

Inspired by silk webs spun by spiders, Microsilk by Bolt Threads is produced with less environmental damage than traditional textile manufacturing. Bolt Threads explains that 'we develop proteins inspired by these natural silks by using bioengineering to put genes into yeast. We produce the protein in large quantities through fermentation, using yeast, sugar, and water. We isolate and purify the silk protein, then spin it into fibres; we knit these fibres into fabrics and garments.'

Fashion giants Adidas By Stella McCartney and Best Made Co. have used Microsilk in their collections, ultimately creating fully biodegradable products. No spiders are used in the making of Microsilk.

Stinging Nettle fibres

Stinging nettles are extremely easy to grow. They can be grown sustainably, organically and need little water. The nettle fibres are naturally antibacterial and mould-resistant; they are versatile and can adapt to different climates. When spun they can be turned into tough textiles for upholstery purposes or into fabrics similar in quality to linen.

What can you do?

Reuse, reduce and recycle! Remember, swapping clothes, renting, upcycling and using recycled fibres are always the most sustainable options.

Look for alternative and sustainable options for all new fabric purchases.

Research your favourite clothing brand and their ethics and sustainability statements (use the Good On You website to do so). Be sure you know where your item is coming from and how the fibre was produced.

Innovate! Do you have the next best alternative to a textile?

SEE ALSO: Agriculture, p. 27; Animals in the Textile Industry, p. 40; Climate Emergency, p. 64; Fashion and Clothing, p. 139; Habitats and Biodiversity, p. 180; Plastics, p. 253; Species Extinction, p. 285

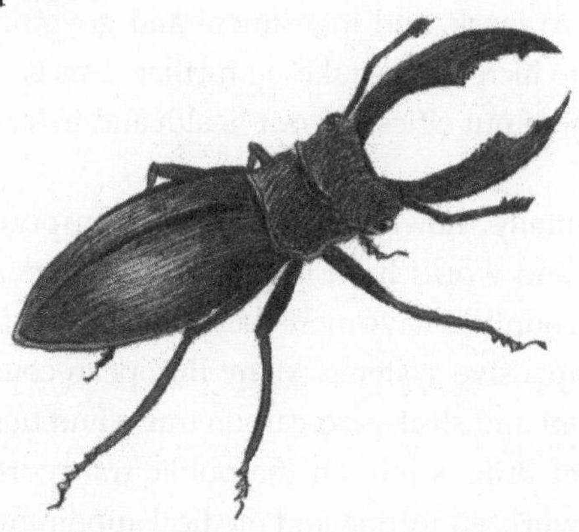

T

TRANSPORT

We are extremely privileged in the UK with the number of transport options we have today, both private and public. You probably groaned reading this. Trains break down. Of course you have to wait for a bus in the cold. Buses are often busy and noisy and sometimes you have to stand the whole way home.

The coronavirus lockdown restricted our use of all forms of transport other than walking and cycling and a national revolution of cycling began. More people then began walking and cycling to work, and investment and government attention helped to increase the take-up further. This is such a good development of our cities, for our health and to lessen air pollution.

Fundamentally, however, the British transport system is useful, safe and works hard to get us where we need to be. Many other countries have more dangerous, dilapidated, fuller and more expensive systems. Many European countries have plentiful silent and sleek zero-carbon trains and buses.

Millions of Britons rely on the public transport system to get them to work, education and medical appointments day in and day out. It is a critical part of our infrastructure. It is a vital service and should be given all government support necessary as part of a forward-thinking zero-carbon future.

However, any transport that runs off anything but your own energy is damaging to the environment in some way, including the carbon dioxide emissions from manufacturing. Your choices could make a big difference to your carbon output. We can travel around by air, water and land transport, which includes railways, roads and off-road transport, such as cable cars.

The most environmentally-friendly way of travelling is under your own steam: walking, running, cycling or scooting! That is anything that involves using your own energy to get to your destination without polluting the planet.

When we are worn out from all that exercise, the next best way we can travel around is by public transport. This is a fuel-efficient way to get around. If you multiply the distance travelled on a litre of fuel by the number of passengers carried, public transport is much more efficient than private transport. For example, it is well known that a car carrying one passenger can travel 9km on a single litre of petrol, but a bus carrying fifty people can travel 50km on one litre each: 50km per litre, per person.

Buses and coaches do pollute and require the road system to be maintained; however their fuel efficiency reduces the environmental damage significantly. Big cities now have hydrogen buses, zero-emission zones, electric vehicles and catalytic convertors on their fleets.

If we all used public transport as a means of travelling, there would be less air pollution and a reduced contribution to the climate emergency. Travelling by public transport is also cheaper than private travel.

The invention of the car at the end of the nineteenth century enabled people to acquire personal motor vehicles. They, and efficient bicycles, replaced the horse and trap. However, they also replaced horse faeces with exhaust fumes, which would later kill many millions of people.

It is not feasible for everyone to always travel in the most environmentally-friendly way. For example, if you have lots of things to carry or if you are a disabled person or elderly and can't manage on public transport you may consider using your own car or a taxi service. London has a zero-carbon fleet of electric taxis which have special charging points at taxi ranks. Renting a car for occasional journeys or holidays allows you to use electric vehicles to good effect. There are also community car share schemes, bike rentals and short-term van hires for moving goods.

Similarly your extended family could live in a different country and you may wish to visit them for life-changing reunions and to consolidate cultural exchanges. People sometimes have to travel when they are sick or to attend funerals of close relatives. Others travel with small children, and even more people have disabilities, seen and unseen.

Many people use their cars as a status symbol: 4x4s, convertibles and faster fuel injections get people excited and exhilarated. Cars are an accepted way of travelling for millions of people, and essential for some. Many people would not be able to do their jobs without them. We need zero-carbon petrol alternatives, electric vehicles maintained by renewable and green energy and sustainable and recycled oils. These alternatives are currently only 10 per cent of the market but that figure is steadily growing.

Vehicles that run on petrol and diesel are the most damaging. They emit carbon monoxide, carbon dioxide, unburned hydrocarbons and oxides of nitrogen. These contribute to both acid rain and the climate emergency.

Air pollution from the exhausts of vehicles burning fossil fuels can lead to illnesses such as lung cancer, bronchitis and asthma. Congestion on the roads creates dangerous and dirty cities and town centres. This all encourages pollution, noise, fumes, illnesses and accidents, and the more roads that have been built, the more vehicles have filled them.

Leaving your car engine running when you are not driving is called idling, and vehicle idling produces up to twice the emissions of a moving vehicle in the same time frame. With this in mind, the UK government introduced laws to discourage drivers from leaving the engine running. You are at risk of receiving a fine of between £20 and £80 if you are caught idling your engine while parked on the road or in a public place. Each minute an idling car produces enough exhaust emissions to fill 150 balloons with harmful chemicals, including cyanide, NO_x and PM2.5s. Stop-start systems in new cars provide a 5 to 7 per cent reduction in carbon dioxide emissions.

Globally, almost 1.35 million people die in road crashes each year, and many more people are injured as found by the World Health Organization (WHO). It can clearly be a dangerous, life or death way to get around.

Illegally parked vehicles encroach on space needed by cyclists and pedestrians. The current Department of Transport favours roads and motorways over cycle lanes, paths and better, cheaper public transport. In 2018, a £28.8 billion government annual fund was announced, to be spent on upgrading and maintaining motorways and other major roads; whereas only £48 billion was designated to cover the next five years for public transport. In 2020, due to the coronavirus, the government announced spending of more than £2 billion on immediate improvements to cycle lanes and pedestrian areas to encourage and maintain the surge in cycling as a safe form of transport.

Better public transport facilities for all of us would mean

better access for elderly people, disabled people and people with children, who all use the system disproportionately more than other social groups. It would mean more attention paid to the safety of trains and passengers, more late-night services in cities, and other improvements.

Although it may take a while for all of us to get used to the idea of walking, cycling and using buses and trains rather than cars, the benefits for us and the planet would be enormous. Healthy hearts, a better mental outlook, more connection to our community and physical exercise healing the whole body are just some of the benefits to us.

Friends of the Earth's long-standing campaign calls for urgent government action to tackle air pollution, which causes 40,000 premature deaths every year in Britain and costs the UK economy £20 billion.

They are campaigning for 'Effective Clean Air Zones', meaning fewer and cleaner cars on our roads, safer streets, more welcoming neighbourhoods and, vitally, healthier lungs for our children.

We need investment in clean, affordable and reliable public transport, alongside an improvement in infrastructure to support alternatives to driving, such as safe cycling and walking.

Everyone appreciates a holiday, some time off from our busy lives. A getaway in the sun is a desirable privilege some may have on a yearly basis. Some may have to travel abroad for work purposes or to visit relatives living abroad. We can travel from country to country quickly and conveniently by aeroplane. However, the exhaust produced from aeroplanes contains dangerous and pernicious air pollutants, including sulphur dioxide and nitrogen oxides. These air pollutants are the fastest-growing source of greenhouse gases driving global climate change, currently accounting for around 5 per cent. In

2010, the aviation industry carried 2.4 billion passengers; in 2050, that number is forecast to rise to 16 billion, according to the WWF. Without action, emissions from increased air travel were forecast to triple by 2050.

The United Nations' civil aviation body has agreed to put a cap on the emissions for an international sector rather than a country, with the aim of curbing international aviation emissions which are increasing at an uncontrollable rate.

Air travel was slowed dramatically in 2020 because of the coronavirus pandemic as planes were grounded in airports and only essential and emergency travel was allowed for months. Airline travel worldwide slumped overnight. People managed to meet online, by email and video link. Millions of journeys were curtailed and people learned new and successful working habits. The air was cleaner.

In the wake of the coronavirus pandemic, the problem of how to avoid spreading infection through air circulation in the plane further stressed the fundamental economics and safety of flying. Investors have withdrawn from conventional airline industries, fleets of aeroplanes have been grounded and companies have laid off tens of thousands of staff. Sustainable air travel is a hot topic and many industries are working hard to find a zero-carbon and energy-efficient way to fly.

What can you do?

Using our own energy is the best way to travel and helps to keep us fit and healthy too. Start today by walking or cycling more and you will feel the benefits quickly.

Use public transport whenever you can (unless you would rather walk or cycle).

Driving an electric car has a negligible impact on lowering your carbon dioxide emissions if the electricity used to charge the battery is generated from fossil fuels; although the proportion of the UK energy supply derived from renewable sources has increased hugely in the last ten years, it is still only 40 per cent at the time of writing.

Go slowly, travel only when absolutely necessary, reduce your air travel and when you do travel by air offset your carbon footprint – support charities that plant trees, and plant your own trees.

Holiday in the country where you live, have a 'staycation' – find beautiful areas where you can relax without the need to fly.

Support Friends of the Earth and other campaign groups by backing their clean air campaigns.

Don't vandalise – or sit by and watch other people vandalise – buses or trains. This takes them out of action for days and causes more problems and delays. (See 'Graffiti' on page 171.)

Decide what kind of public transport you would like to see and then ask questions, lobby, write and give feedback to your local provider.

You may want to join the Campaign for Better Transport, that campaigns for transport efficiency.

Follow and support the following organisations and charities that are campaigning for and working towards better and safer public transport: Action For Rail, Bus Users, Greener Journeys, Living Streets, Road Peace, Sustrans and We Own It.

SEE ALSO: Air Pollution, p. 29; Climate Emergency, p. 64; Cycling, p. 92; Holidays, Travelling and Relaxing, p. 191; One Planet Living, p. 234

TREES AND TREE PLANTING

The single biggest and most positive action you can undertake to support the planet is to plant a tree. They produce oxygen while sucking up carbon dioxide, the main contributor to global heating. Four big trees can absorb about a tonne of carbon dioxide. In return we can breathe!

A tree absorbs the carbon and at the same time creates a cooling effect for the land around it. A tree is the foundation and home for millions of insects, birds and animals. A tree offers shade and cool air on a hot day for humans and creatures and a safe night-time environment for birds and insects too.

A tree lives for hundreds of years and offers many generations of creatures a safe home. Trees have nuts, fruits, leaves and bark that can be eaten, made into flour, flavourings, tinctures and medicines.

When trees are introduced and planted in a city, the overall temperature falls and local people have shade and cool air at street level in the hottest summer months. Some governments have introduced schemes that plant fruit-bearing trees so local people can have free food in abundance from trees. The People's Trust for Endangered Species will give grants for people to replant their old and ancient apple orchards.

When trees are planted in desert areas, they help to bring a stable ecosystem into the soil. They fix nitrogen in the soil, hold the soil in place, and allow other species to grow around them successfully. Water is created and attracted to places that have abundant trees, and streams and rivers flourish in woodland areas.

Trees have been sacred in religion for thousands of years. The British Museum has pictures of the gods and goddesses being part of a sacred tree in religious imagery.

Our native trees were honoured and revered by the Druids, the spiritual religious organisation of Britain from around 1,000 BC to AD 150. Druid lore, myths, rituals and lifestyle centred around the oak tree as the most sacred of all living plants on earth. The tree's acorns were an important source of food at that time and a staple carbohydrate. Yew trees are also considered sacred and many are found in churchyards across Britain, as churches were built on land that was once a sacred grove of trees.

Tree dressing and decorating has been a part of our community rituals for a long time. People tie ribbons on a tree to signify the loss of a loved one or to make a special wish. In our culture tree dressing was an ancient act, used for many community events, and is a ritual way to honour our dead and to make resolutions with nature with the tree as the witness. Common Ground reinvented tree dressing as a way to love a favourite tree and to show that every tree counts. Their Tree Dressing Day takes place on the first weekend of December. Of course, we still 'dress' trees for the ancient festival of Yule. Now we call it Christmas.

Apple trees are also very important in British history. There are more than 3,000 varieties of apple in Britain! Singing or 'wassailing' to the apple tree to encourage its fertility was a part of the British folk tradition for many hundreds of years. You can still sing to the apple trees in Devon and Cornwall in January and February each year, which locals do to encourage the apples to produce wonderful cider for them to enjoy.

Walking in a forest of trees is a healing and calming thing to do. A person who is mentally or emotionally unwell or unbalanced can find extraordinary calm and nurturing from a day spent in the forest. 'Forest Bathing' is known as a health remedy across the world.

Planting trees can help mitigate the negative effects of your lifestyle in some cases. So, for example, if each passenger on a flight from London to New York were to plant eleven trees, they would mitigate the environmental damage caused by the carbon dioxide produced by their journey. This is carbon balancing, albeit on a long timescale. Scientists estimate we would need to cover a quarter of the earth in trees to rebalance all the aeroplane journeys we have collectively taken.

The National Trust announced in January 2020 that the organisation is planting a fantastic twenty million trees in Britain during the next ten years. The direct aim is to mitigate climate change. This is the equivalent of planting forty-two Sherwood Forests and is part of the organisation's 125th birthday celebrations.

What can you do?

Plant trees!

It is the most powerful physical thing you can do for the planet and the solution with the longest time frame for your lifetime.

You can plant trees for newborns, birthdays, celebrations, weddings, deaths; every single time you and your family celebrate something use a tree-planting ceremony to ecologically mark the occasion.

Use a reputable charity that already has land and tree-planting experience if you can. They will have a high success rate, with the trees more likely to survive. Your trees may not survive if you don't take specific care planting and watering seedlings and young saplings in their earliest months.

Plant species that are native to your country and soil such as the oak in the UK. Don't plant trees and bushes that could cause a problem later. Native species will always be the best ecological option.

Plant trees with plenty of space for the roots to spread; the tree's chances of survival increase if you dig it a square rather than circular hole. The roots colonise the soil around them and spread properly, giving the tree optimum chances to thrive. Circular holes make the root system develop in a spiral and weaken the roots.

Leave trees alone. When a tree is damaged after a storm the trunk is often cleared away or burned. If you leave the tree to fall and rot gracefully on the spot it can still be an incredible ecosystem for species to continue for many years to come.

Learn the secrets of the trees. How to listen to them, how to sensitise yourself to their needs and wants and be aware of their requirements. Ask a tree where it wants to be planted, ask it when it wants watering.

If you haven't got land or a garden to plant trees you can adopt one in your local street. You can water local trees when the weather is very hot. You can name them and dress them for fun, for festivals and events.

Organisations that work with trees and tree-planting include: Common Ground, The Woodland Trust, Woodcraft Folk, The Tree Council and TreeSisters.

SEE ALSO: Climate Emergency, p. 64; Habitats and Biodiversity, p. 180; Species Extinction, p. 285; Tropical Rainforests, p. 304

TROPICAL RAINFORESTS

In Indonesia and Borneo and other countries situated on the equatorial line, the deforestation of rainforests is a serious problem.

Palm oil is the most widely used and ubiquitous vegetable oil in the world and its production is the largest cause of rainforest destruction. The demand for palm oil is so great that it has destroyed the integrity of the rainforests in Indonesia and Malaysia, where production accounts for 85 per cent of the world's palm oil.

Palm oil is found in our supermarkets in more than half of all packaged food as well as in cosmetics, soap, detergents, toothpaste and a wide number of household products. When you see 'vegetable oil' listed as an ingredient, it will almost certainly be palm oil.

To grow oil palms the companies need high humidity and a consistent temperature and a wide planting area. It has been estimated by Greenpeace that 300 football fields of rainforest are being destroyed every hour for palm oil production.

When the forests are cleared, species that lived in them become stressed and pushed to extinction, such as the Indonesian Orangutan, the elephant and the Sumatran tiger. Companies are more committed to sustainable palm oil production after a successful campaign by Greenpeace started in 2010. The Consumer Goods Forum has been working towards sustainable palm oil. However, the Indonesian government recently announced plans to cut another 18 million hectares of rainforest and create more oil palm plantations.

Destroying the rainforest exacerbates climate change. Rainforests are the earth's largest carbon sink, storing the greenhouse gases that have been causing climate change. Every time they are destroyed incredible quantities of carbon dioxide are released into the atmosphere. Indonesia is the third biggest contributor to the planet's CO_2 because of its destructive rainforest policies.

Indigenous people are affected. An oil palm plantation destroys the forest, the natural and clean water and all the forest resources that indigenous people rely on. Many face poverty for the first time because of rainforest clearances.

Huge companies like Cargill are accused of not doing enough to stem the destruction of the forests, reducing biodiversity and harming the indigenous people who have lived in the forest for generations. The Rainforest Action Network is just one group trying to work with the palm oil producers to create a more sustainable industry.

Logging for timber is another major threat to the world's tropical forests. Paul Pavol, an auto parts salesman from Papua New Guinea, became an anti-logging campaigner and won a US environmental award in 2018. He spoke out to raise awareness about his country's environmental policies: 'For many generations, my community has farmed, fished and hunted in the forests of Pomio, all while keeping them healthy and standing. In only a few years, Rimbunan Hijau [the world's largest tropical logging company] has clear cut many of these forests, leaving behind a desert in their place. As a result, we're starting to lose our customs, our culture and our traditions. We know that this is an uphill battle, but fighting is the chance we have to hold on to our way of life.'

What can you do?

Join groups and charities campaigning to educate the public about the terrible effects of mono crops and palm oil plantations in Asia and Africa, including Greenpeace and the Rainforest Action Network.

Try to avoid processed and packaged foods – they are very likely to contain palm oil. Read the ingredients list on pre-packaged foods. Avoid products and furniture made from tropical hardwoods. Many wood and furniture retailers have information about their sustainable efforts and indicators.

Get familiar with the other names palm oil can be listed as: for example, palm kernel, palmate, vegetable fat and many more. Write to companies that claim the palm oil they are using in products is sustainably sourced. Ask them where the palm oil comes from and what makes it sustainable.

Look out for the 'palm oil free', 'certified sustainable palm oil' and 'FairPalm' logos displayed on products.

Stick to real foods like organic fruit and vegetables and unprocessed items to be sure that you are not contributing to the palm oil problem.

Support organisations and charities working to protect our rainforests around the world, such as: Rainforest Action Network, Rainforest Alliance, Orangutan Foundation and the Orangutan Care Centre.

SEE ALSO: Agriculture, p. 27; Climate Emergency, p. 64; Fast Food, Takeaways and Restaurants, p. 144; Habitats and Biodiversity, p. 180; Species Extinction, p. 285

V

VEGANISM, VEGETARIANISM AND THE 'ISMS' OF OUR FOOD

Eating food is complicated. We all have to eat, every day. The politics of food, as you can see from those issues we have touched on in this book, is multifaceted, complex and interconnected.

A large proportion of the people on the planet have chosen not to eat meat, opting for a diet of grains, vegetables and fruit. Proteins that the body needs for survival come from a range of sources, such as pulses, cheeses, tofu and fermented foods. They have a much smaller ecological footprint than a family who eat red meats every day.

What you eat depends on where you are and the range of local foods around you. Native foods all have the perfect ingredients and essential minerals needed for where you live, for example, vitamins or minerals contained in local water, vegetables or fruits. Local food won't have travelled far. Local food will have all the nutrients you need to survive the terrain, the weather, the ecosystem and its challenges.

On our coastline in Britain our wintry months reveal the

perfect medicine in sea buckthorn; like tiny golden pearls, nature allows them to flourish in the coldest places. We may also harvest the rosehips and the red berries of the hawthorn to supply hot drinks and vitamin C during the autumn.

Every ecosystem has its special food and medicine, where the local people understand and benefit from what grows around them first.

For centuries Welsh people have eaten a simple soup that they call *cawl*. It is a soup consisting of all the vegetables of the day, usually onions, leeks and potatoes from the household vegetable patch, and leftover scraps of meat which are usually lamb. This is a supper which has been put into millions of bowls over many generations. Family recipes for the dish are shared for generations. Many people grow their own vegetables or use the local market system that has survived all economic pressure by other businesses.

This is a local food and the people do not call themselves environmentalists, or some other label. Rather they eat simply with little waste, no packaging or manufacturing, negligible food miles and a clear knowledge of where their food comes from; but they also eat meat.

Human beings have eaten other animals for thousands of years. Eight thousand years ago in the Mesolithic period of Britain's history we know that a spiritual communion went on between animals and humans. We can see they ate a very varied diet of meat and fish. We can see the red deer masks in the British Museum from Star Carr in North Yorkshire proving group animal and human ritual.

This spiritual communion can also be seen in modern indigenous people's rituals. A sacred prayer begins the journey between the hunter and the hunted. Many observances and taboos were and still are involved before a hunt takes place; there is a spiritual link to the killing of the animal that is sacred and complex. When an animal is killed and gratitude is given not just to the life of the animal but to the Creatrix, the goddess who regenerated the animal kingdom in a state of fertile abundance in this way, the killing is deemed honourable. This kind of ritual observance with the hunted animals still happens in indigenous tribes today.

This sacred relationship between what we eat and how we eat it continues in many forms. It is quite clear that the food and lifestyles of each of the thousands of tribes and clans and families around the world contain remnants of this form of gratitude. Some say 'grace' or give thanks before a meal, especially at important meals, occasions or festivals with family and friends and loved ones present.

When this awareness and consciousness is part of your daily routine you will choose food that is good for your body and right for who you are and the conditions in which you live. The food that grows in your own country and the animals that live around you, if you eat meat, are usually the kind of foods you should eat.

For many people the multitude of political, environmental and societal considerations about food are complicated by guilt, shame, confusion and even hunger.

Choosing to live a life without the death of animals and insects or the inadvertent use of animal by-products is complicated and potentially difficult. Those that have transitioned to a fully vegetarian or vegan diet have often reported feeling better in their bodies and in their heart. We all have to eat more vegetables.

An increasing number of people are helping the planet by reducing their consumption and their carbon output, and lowering their blood pressure by reducing their meat and cheese intake. The World Health Organization (WHO) has told us that most people in the West need to eat a lot less red meat, less complicated food, less spicy and more simple meals. We need to do this for our health and for the planet. WHO remind us that the world's leading cause of death is heart disease and cancer, known to be directly linked to red meat consumption. Our body's reaction to excess meat consumption and over-spicy complex food also causes or exacerbates other health problems like heartburn and ulcers.

Environmentally it's of no value to swap one product for another if it causes the same amount of damage or pain. Awareness of what's right for your body is more important. Awareness of the amount of food that you eat consciously rather than unconsciously is what's important. We need to say this because there is no one right answer.

There are plenty of indigenous tribes who survive by eating meat but leave no trace and live perfectly well without creating waste or carbon pollution of any kind. There are plenty of others who claim a moral high ground as vegans but eat avocados every day as if this fruity jewel is available to everyone, despite the fact that its growth as a vegan alternative is causing droughts and local people to lose their drinking water supply in another country. See *Rotten: The Avocado War* (Netflix, 2019).

This judgement of people by the food they eat is causing emotional shame and complicating our relationship with food. Many people are made ill by food, every day. Many people are so ill they can barely eat anything at all. Other people have no access to good food, whether it be a garden to grow things in or a food bank. When food is scarce human beings do extraordinary things to survive.

There are more people on the earth hungry today than ever before. They wait for the rain for their crops to grow. Millions of people on earth live like this. The food system is an agricultural business, and the majority of people on earth fall below the income levels required to buy things from this agricultural system. They grow what they can and eat when there is food on the table.

It is not necessary for everyone on earth to be a vegetarian or vegan to save the planet. It is necessary for people to become more conscious. Those that do change the way that they eat and understand the world have taken a personal step towards a more compassionate lifestyle. It requires four times the amount of land for a cow to produce meat than it does to produce a wide and interesting variety of vegetables, roots, leaves and bulbs.

It makes sense for the body to eat a variety of fruit and vegetables as the core of anyone's diet. Some people who eat a meat-based diet have been found not to eat vegetables at all. A diet concentrating on red meat especially is known to trigger a variety of cancers.

It is neither sustainable nor biologically possible for everyone on earth to eat meat. It is neither sustainable nor biologically possible for everyone on earth to eat avocados, cashew nuts, almonds or other exotic fruits and nuts too. Can you see that everything you eat creates a cause-and-effect?

Many people see themselves through the identity of their food.

When love makes a hot soup, those that eat that soup know and feel the nurturing inside, however simple.

When the identity of the food is about love, the message transcends all our judgements and quibbling over its gastronomic impact or its environmental footprint. Simple food brought from the land surrounding you, from local growers, and cooked with love is the best solution.

Food has made us ill, from the chemicals and pesticides inside it to the emotional breakdown and the complications of over-eating and obesity, eating the wrong food or not wanting to eat food at all.

The 'isms' of vegetarianism and veganism are currently popular, and some people think it's a club that everyone must join. The complications and the politics of food production make that impossible in today's climate, even when some meats are more damaging than others.

Is it more important to stop buying vegetables that have been flown 30,000 miles in an aeroplane and then kept in a large refrigerated container so that you can eat them fresh all year long? Is it more important to stop expecting exotic nuts or fruits to be available every day in your diet like a queen or king in ancient times, even if the place where they are grown has no water left for the villages and people that grew them?

You can't have it all; and the planet doesn't have enough. Your plate must be filled with vegetables and foods and conscious proteins. Your hunger must be sated first from what is grown around you, not from the exotic foods of another culture. The foods of other cultures are to be eaten as special and extraordinary things, enjoyed as a rarity rather than an everyday occurrence.

What can you do?

The Western world has a collective food neurosis and has created a complex emotional relationship with food. Eating is something humans do every day and yet we rarely stop to offer gratitude for our food or the enjoyment and pleasure that it brings to our lives.

Take a moment and say thanks to the earth! Everything that you have eaten since you were born has come from the earth. Touch the earth and connect with gratitude.

From this simple place where your heart connects with the heart of the earth and acknowledges the abundance that the earth has shown you, you can start a relationship with the earth herself. From here you can start a relationship that is connected and real with the fruits and vegetables that you touch and acknowledge them as your fuel, your comfort, your dinner.

Food choices create stress in people and stress itself causes illness. Forgive yourself for everything you've eaten in the past that was not right for you and resolve somehow to be more conscious and sensitive now to what it is that you're eating and how it makes you feel and the environmental impact that it has on the planet.

Don't judge someone else's diet, food habits, cravings, addictions or choices. You have no idea what is happening to them.

Don't force anyone else to make the same dietary choices as you unless they want to.

Make beautiful food with love and everyone will want to eat it.

The following organisations provide information and support: Meat Free Mondays, The Vegan Society, The Vegetarian Society, World Meat Free Week, Counting Animals, Mercy for Animals, Food Empowerment Project, Compassion in World Farming, Animal Outlook.

SEE ALSO: Agriculture, p. 27; Climate Emergency, p. 64; Consumption and Consumerism, p. 77; Fast Food, Takeaways and Restaurants, p. 144; Food Waste, p. 156; Your Carbon Footprint, p. 322

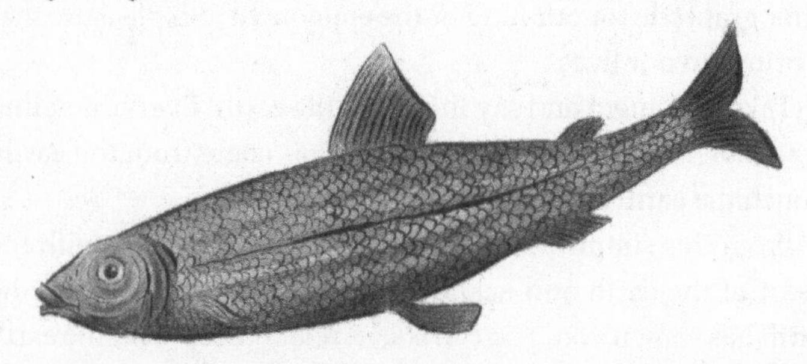

WALKING

Of all the empowering things you can do for yourself and the planet, if you are able, a love of walking would be very high on the list. Walking reduces stress, helps you attune better to nature, helps you see your local environment more clearly, helps you to connect with the world and keeps you healthy.

Walking allows you to hear the sounds of birds and trees or crunchy leaves underfoot. Walking allows your feet to connect with the earth and your body to slow down.

When you walk, you don't need a lift, to wait for a train, or to learn to ride a bike. No money is involved. You are on your own two feet and can change your mind, turn around, stop and move elsewhere under your own power.

Most daily journeys are less than 3km in the UK. In fact, more than 40 per cent of all car journeys taken are less than 3km and in London that goes up to 60 per cent of all journeys. That is an easy distance to walk for an able-bodied person. The first five minutes of a car journey actually causes more pollution than the rest of the journey and therefore short car journeys are adding enormously to the pollution problem.

Motorists are exposed to high levels of polluting emissions at the start of their journey and therefore there is an increased health risk to the driver as well.

A recent London marathon found that as cars and local drivers were restricted during the event, actual measurable pollution levels in the capital dropped dramatically – by 89 per cent. This has also been seen in every city around the world that has had a coronavirus lockdown. The improvement in air quality within just a few weeks has been miraculous. People started walking for health again in parks and by rivers or in areas of nature. Walking became very popular and necessary.

What can you do?

If you want to nurture your body, walk in nature. If you want to heal your mind and calm your heart take a walk and connect with something that is alive: the local river, woodland or wild path.

Walking on the earth regenerates you and rejuvenates you. Barefoot on the beach, in the grass or at the side of the river; your feet connecting to the earth reconnects you as a healing process.

Being in awe of breathtaking scenery, birdsong, the sound of oceans and rivers also stimulates endorphins. Standing on a cliff edge with the sun setting in glorious colour can elevate your feelings. You can literally cheer yourself up by walking in nature and allowing your senses to experience the rawest and purest form of emotional medicine.

Start walking more, simple. Walk first.

Make a plan of your activities and week-to-week appointments. Work out when you can walk or walk part way to increase your daily steps. See if you can get friends involved for support.

Walk to work. If you are a young person, save your parents a car journey by walking to school or college. Get the bus if you can't walk. Walk first, then bike, bus, train, coach or even boat. This is the green transport hierarchy.

Walk down and upstairs instead of taking the lift. It saves energy, and it gets you healthy.

Various medical and health organisations say that a person should walk between 6,000 and 8,000 steps a day; many people take a challenge to walk 10,000 steps a day for better health.

SEE ALSO: Climate Emergency, p. 64; Cycling, p. 92; Gaia, p. 165; Transport, p. 294; Your Carbon Footprint, p. 322

WATER

We are water; 60 per cent of our bodies is made of water. Our brains are composed of more than 73 per cent water, our lungs 83 per cent. A baby has more than 75 per cent water in its body. We need fresh drinking water to survive.

The earth is also mostly made of water. 71 per cent of the earth's mass is water found in the oceans and seas, but about 96.5 per cent of that is sea water – salty, and therefore not suitable for human consumption. Of the remaining amount of fresh water, only about 1 per cent is available for drinking. Much of the rest is trapped in icebergs and glaciers.

The water you drink has been on earth a long time – hundreds of millions of years, in fact. It is all recycled around the earth and back into our bodies again.

Population increase has meant that although the amount of water available is the same, there are more people that need it every day to survive. Therefore clean, fresh drinking water has become a rare and special resource. Many countries have problems supplying fresh clean water to their citizens.

Lack of clean drinking water has been the reason for the flight of so many environmental refugees: people have to leave their homes if the water runs dry.

In rich countries clean drinking water is often used for cleaning, for flushing toilets and for other industrial and business uses. It flows freely out of a tap, giving the impression that it is plentiful and abundant.

Water is needed in huge amounts to allow our lifestyle to continue. The *Encyclopedia of Water Science* has calculated how much hidden water is used for everyday food products. The production of the average hamburger, from beef to plate for example, needs 2,400 litres of water.

Avocados, special fruits and nuts, for example almonds, are crops that need disproportionate amounts of water compared to more basic crops. These industrial fruit crops exacerbate drought conditions and contribute to the drying of rivers and local water supplies.

In Chile many villages in the Petorca region, the largest avocado producing area, have been complaining that their water rights have been violated. Illegal water irrigation by avocado farmers growing directly for British supermarkets has caused a severe water crisis for locals.

Local people now have no access to clean water and must use drinking water brought in by truck.

In other countries clean water is scarce. Sometimes water has created wars it is so precious. The United Nations says that nearly two billion people live in regions without proper access to clean water.

It is a critical element for the survival of every species. Without water you will die, animals and insects will die. Without water, plants and trees cannot survive. Without water, the land is grey and lifeless, or burnt, orange and golden sand, dry as dust.

Many indigenous people are fighting for water justice and access to clean water. They say water is sacred and it has been recognised as such by their ancestors since time began. Some people believe water represents the blood of the goddess of the earth as is seen in indigenous narratives from North America, and Mami Wata is the African deity or spirit of water for the entire continent. #WaterIsLife

In 2010 the United Nations General Assembly declared that access to safe, clean water and safe hygienic sanitation was a basic human right. Today, 785 million people do not have access to clean water close to their homes; two billion people don't have a decent toilet in their house; 31 per cent of schools worldwide don't even have clean water.

In some countries water is scarce, expensive and controlled by corporations who bottle it and sell it to consumers with a mark-up of 60 to 200 per cent according to Business Insider. When water is controlled by corporations and corrupt governments, people suffer and lack this most basic right.

People need about 2–3 litres of water a day for drinking and in food, and another 45–65 litres per day for washing and general use; the figure is even higher when we eat out and use outside resources. By taking shorter showers and turning off the water whilst you lather, you can save 17 litres for every minute saved; you can also save 6 litres of water by simply turning off the tap when you are brushing your teeth, as found by Friends of the Earth.

Water is becoming scarce. It is estimated that by 2050 there will be a 22 per cent shortfall in the water needed because of the climate emergency.

Every minute of every day a newborn baby dies because it doesn't have clean water and because of the complications that brings. Another child under the age of five dies every two minutes from diarrhoea due to drinking unsanitary water. Good hygiene and clean water conditions are vital for basic health and survival.

In 2015 the Georgie Badiel Foundation was set up by model and environmentalist Georgie Badiel. 'As a young girl, I used to walk for three hours every day to fetch water. After travelling the world, I went to visit my sister who was almost nine months pregnant. She still needed to awaken between 2 a.m. and 4 a.m. to get water. This is what inspired me to start making a difference to the critical need for water and sanitation.' The foundation has since made clean water accessible to over 100,000 people in her home country of Burkina Faso by building five wells. The foundation has also trained and educated fifty women to restore, maintain and service their villages' wells.

We are water. The water that is outside us in the environment is a reflection of the water that is inside us. After all, it's all been recycled in our bodies. We are water.

Everyone has a fundamental right to drink clean water. The complications of poor government and planning, and industrial greed, make it difficult to provide clean water for everyone to drink.

When our leaders wholly concentrate on the importance and imperative of the continuing cycle of clean, clear water they can see it is a never-ending self-sustaining system. Human interference and greed have taken too much in places where it is needed. Water and its strategic and life-giving properties mean it should be at the forefront of every government's environmental agenda.

What can you do?

Conservation of water is important and worthwhile. Treat water as an expensive, precious and life-affirming liquid because it really is. Save water, everywhere, all the time.

Don't leave the tap running. Have a shower instead of a bath.

Carry out a ritual of thanks for water, both in your body and on the earth.

Reuse water especially in dry summers. Use a rainwater collector all year long.

Don't use a water hose in the garden unless you really have to; let the rain do the job first.

Contact some of the water and campaigning charities and pressure groups to see what you can do to join them and help or publicise their issues in your school, college or workplace. Be aware of toxic chemicals and microplastics in products that you may be pouring down the drain.

WaterAid is an international charity that has been successful in bringing clean water, decent toilets and safe toilet hygiene across the world. They have reached 27 million people with clean water so far.

Campaigns for clean, accessible water include: Water UK, Refill UK, Give Me Tap, Frank Water, Consumer Council for Water, the Love Water campaign.

#showwatersomelove #WaterisLife

SEE ALSO: Litter, Waste and Landfill, p. 218; Seals and River Life, p. 276; Seas, Oceans and Rivers, p. 278

Y

YOUR CARBON FOOTPRINT

> 'You have to listen to the science and the facts because climate change isn't an opinion.'
>
> Alexandria Villaseñor, organiser of FridaysForFuture New York, and founder of Earth Uprising, May 2019

Everything you buy has an environmental and ethical price – not just the price you pay at the till. There is the price that nature paid for the raw material to be mined or extracted and the process for each item to be made. There is the price the workers and designers paid for their labour, from farming to sewing; human hands also made what we need.

There is a price to be paid for the consumption, the shops, the adverts; the money flows in and around our economic system. There is a price for its disposal and final resting place on earth.

The environmental price is barely counted. The carbon footprint calculator attempts a numeric value in terms of the pollution output. It is a crude and negative calculator, but until we have another means of knowing, it is the current favoured tool.

We need a new calculator to show us the wealth of the earth as a positive addition to our society and our human development and not a negative one.

Attributing pollution and environmental damage to be measured by carbon dioxide equivalence is deceptive. Your personal ecological footprint is complicated, and no one walks the same path. The measurements are then given as averages. Not every kind of pollution can be 'carbonised' and neatly added to the data.

This measurement is given as a stark reminder that there is a lot of work to do. This is not a tinkering situation. This is not something that doing the recycling will solve. We also realise that governments and corporations must be involved. A family can do a lot, but a nation working in unison can do so much more.

Most people should look first at where their money goes. Cars, aeroplane travel, white goods, electricity, fashion and beauty; these are the huge household-expenditure items where you can really make a difference in your carbon footprint. Fashion and clothing are big carbon emitters; if you eradicate them from your shopping list you will reduce your overall levels.

The average person in the UK is responsible for 13 tonnes of carbon dioxide production (equivalent) per year according to the WWF. If you want to live within the planet's means (see 'One Planet Living' on page 234), you need to reduce that to 3 tonnes of carbon dioxide production (equivalent) per person per year. That's a dramatic shift.

At the moment we would need four Planet Earths to provide the lifestyle we have in the UK – the same access to resources and energy – for everyone on earth. The rest of the humans on earth can't 'catch up' with us, or 'develop' like us. We are wasteful and use up too much of what there is. There is, after all, only one earth.

Each year we are faced with Earth Overshoot Day; this day is marked when humanity's consumption has exceeded the ecological resource that the earth can generate in that year. Earth Overshoot Day is calculated by the Global Footprint Network, who have reported that over the last few decades, the overshoot date has been falling earlier in the year. In 2019, Earth Overshoot Day fell on 29 July. In 2020 it was on 22 August, a little later in the year due mainly to the reduced pollution and use of resources from coronavirus closing so much of society down. Once Earth Overshoot Day has been reached, we begin to exhaust the planet of the capable resources of that year.

We need a huge, immediate and fundamental shift in the way we live and do business in order to survive and to be sustainable.

Your lifestyle creates a cost; everything has a consequence. We don't want to cancel our education or stop getting healthcare, obviously. We need to use intelligent technology and efficient ways of working. Scientists and inventors have to work constantly to find a way to upgrade immediately so we can survive with nature intact and live a fully zero-carbon lifestyle.

This is the average carbon footprint for a working person in the UK. It is divided roughly like this:

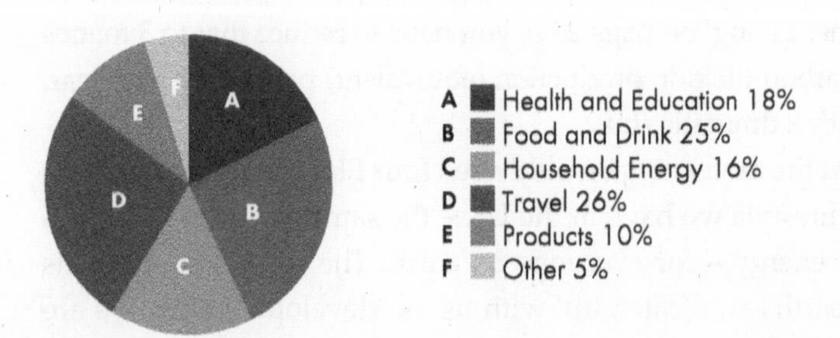

Source: Sir David Attenborough's BBC documentary *Climate Change: The Facts*, first screened 18 April 2019

One shared earth is all we have. According to the organisation The World Counts, we need an overall average of 1.76 Planet Earths to provide resources and absorb the waste we have produced right now in 2020 and the number keeps growing. That means millions of people are living simple, subsistence lives right now using well below one planet's worth of resources while we in the West exploit our wealth and privilege and use multiples' worth of resources.

One Planet Earth is all we have but humanity has not shared the earth's bounty with each other. Rather it is wasteful and greedy, creating systems that favour cheating and hoarding. These are traits that our leaders should not promote. The systems that allowed that have clearly caused us to reach the cliff edge in terms of our survival.

All the other creatures live without the immense trappings, collections and junk accumulated in a lifetime by human beings. They make their nests, burrows and dwellings from natural things. Their footprint can't be seen; only their bones indicate their previous existence.

If your plan is to reduce your footprint you can use all the help you can get. This book will give you hundreds of ideas. Governments and industry must, however, help us all shift the massive impact of road building, stop economic support for the burning of fossil fuels, start promoting energy efficiency, encourage small-scale and local farming and awareness, and support environmental legislation on so many other national and international issues that will make the biggest difference. Then we will begin a serious zero-carbon future together.

What can you do?

Now that systems and calculators have been created you can assess your own environmental footprint, your ecological lifestyle and learn how to change, to upgrade, to downsize or to stop an activity altogether.

Where are the places where depth and meaning are added to your life without adding at all to the pollution on the planet?

How can you assess your life to calculate its success in terms of friendships and love? How can you make a difference? These are the calculations you can make for yourself. These are the calculations that matter when you are ill, facing death or loss.

Envision a future of zero carbon. Many European cities are well on the way to achieving this. We have the technology; we just need the political will and vision.

Take action to reduce your footprint. You can see why you need to do certain things straight away to bring your carbon down.

Adopt energy efficiency at home, turning off energy and enhancing your house with renewables such as solar. Switch your electricity provider to renewable energy or an alternative energy provider; this will make a huge difference to your overall carbon output. There are plenty of 100 per cent renewable companies offering excellent prices.

Reduce fuel use in travel. Walk or cycle short journeys; take the train for longer journeys. Fly only in an emergency – use rail and sail instead.

Be less wasteful in every area of your life. Stop buying new things. Accept gifts with greater gratitude.

Plant trees at every birthday, birth or death, or fund more through charities and groups.

Reduce food intake from abroad. Eat everything you buy. Avoid air-freighted food. Reduce meat and dairy, especially beef and lamb and cheddar cheese.

The average person could reduce their carbon footprint by two full tonnes if they undertook some of these changes.

Calculate the exact footprint for you and your family using one of the online calculators: Conservation International Footprint Calculator, WWF Footprint Calculator, Carbonfootprint.com, United Nations footprint calculator.

Follow organisations, principles and information for One Planet Living. Get informed then make a plan!

SEE ALSO: Climate Emergency, p. 64; One Planet Living, p. 234

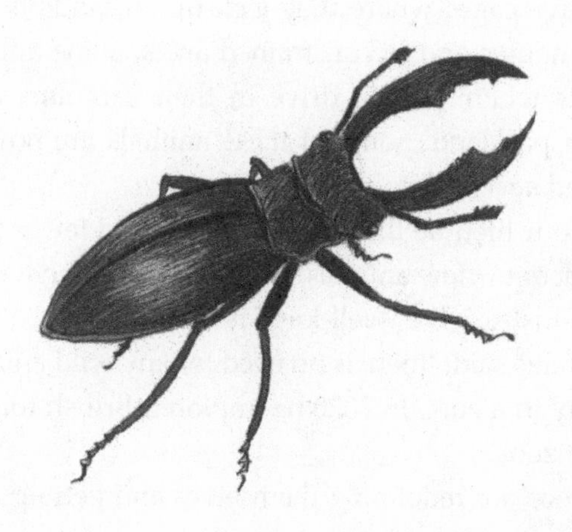

Z

ZOOS AND CAPTIVE ANIMALS

Zoos and wildlife parks come in many types. There are over ten thousand zoos worldwide, five thousand of them in Europe.

Most have cages where they lock up all kinds of animals behind windows and in constrained areas. Some are in more open areas where people drive in their cars and view the animals in parkland. Some of these animals are now on the endangered species list.

Within our lifetime there will be fewer and fewer zoos and similar places to view animals. Even people who have worked in the industry, like well-known conservationist Damian Aspinall, have said, 'there is no need for any wild animal to be in captivity in a zoo'. In 2020 he implored British tourists not to support zoos.

Some zoos are redefining themselves and getting involved in preserving species, even helping them to produce young in a bid to save the species. Biological research is one of the better reasons to support zoos. However, critics think that zoos do not play a big part in conservation and that they have exaggerated their research abilities.

Other public places use animals for show, for entertainment and humour. Think about the times you have seen a reindeer at Santa's Grotto in your town? Have you seen a dolphin show, or witnessed an owl or bird display? Animals are also used on television shows.

The good news is that as of January 2020 circuses performing in England will no longer be allowed to use wild animals as part of their act thanks to the Wild Animals in Circuses Act 2019. That is true in Northern Ireland, Ireland and Scotland. There are 27 countries that have now banned the use of live and wild animals in circuses.

Modern German circuses have started to use holograms of animals instead of live creatures.

At a famous tower, scene of macabre hangings and beheadings in London's previous history, there are still six or seven remaining prisoners. They and their relatives have been captured continuously for five hundred years. They keep their partners for life. These creatures are the last remaining prisoners of the Tower of London.

Their wings are clipped so they can no longer fly. They are the Ravens of the Tower of London. Brought in by Charles II to create a dreadful atmosphere when people were beheaded, these huge birds are now kept as a tourist spectacle. They often try to escape, but their home is small – a metal cage – and they cannot fly away. They have a Ravenmaster who feeds them and makes them photographically acceptable to the millions of visitors from across the world. Superstitions about the kingdom 'falling' if they are set free have turned out to be Victorian nonsense. Yet they are still incarcerated, the last living prisoners of a cruel and barbaric bygone era in British history.

What can you do?

Write to the Ravenmaster, the Tower of London, London, EC3N 4AB, and ask for the unconditional release of the ravens. The RSPB offer help with bird freedom campaigns across the country.

Contact Zoocheck and Freedom for Animals whose mission is to free all captive animals in the UK. PETA also campaign for the freedom of animals.

Don't support zoos, wildlife parks, or other places where animals are locked up.

Pray for animals or envision their freedom. Mourn their loss and the cruel lives they endure because of our sport, our greed and need for spectacle.

SEE ALSO: Ethical Living, p. 135; Gaia, p. 165; Species Extinction, p. 285

OUR GREEN JOURNEY

A Letter to Greta Thunberg

When Bethan Stewart James asked me if I could send her beautifully handwritten letter to Greta Thunberg, I read it with a sense of both pride and sadness. Her letter struck me as a meaningful snapshot of the confusion, worry and determination that so many young people feel today about the planet and its problems. Bethan, like so many young people, has been inspired by Greta Thunberg to challenge the status quo and do something positive for the planet.

This book is the result of that letter and our attempt to clarify her questions, to stimulate debate, to raise ethical awareness and most of all to encourage changes in our actions and our lifestyles. Bethan's Schools Manifesto is her inspiring list of what needs to be done, in her school and her community. We all could create a list of our own, and work with others to make it happen.

Dear Greta, bore da (good morning)

I am writing to you to tell you about how inspired I am because of you and everyone who went to the school strike. It is very important that people notice what is happening on the planet.

My name is Bethan. I am nine years old. I live in a beautiful little town in Wales called Fishguard but soon it will not be so beautiful.

It's not too bad in Fishguard – there is only a bit of litter, but that's the thing: we think it's okay, so we carry on and a bit later the beauty-filled-ness is gone and it's just plastic. That's one of my worries. Let's start off when I was about five – my brother came to the breakfast table and said, 'Did you know? There is an island of just plastic in the sea.' I thought he was making it up.

But about a year later I found out it was real. So I did some more research into it and then I became one of the many people fighting to save our world. Me and my friends are planning on making a litter picking day. I think we should get a bottle recycling machine that gives you your money back like they do in Scotland.

About a year later my brother said, 'All the icebergs are melting.' I knew he was telling the truth this time. Then I found out about cars and fossil fuels and I stopped buying so much after my mum told me that that was bad too and before I buy anything I think, Do I want it? Do I need it?

Last term me and my friends started a club in school called The Litter Picking Club. In our neighbouring town of Goodwick there is a building called the Ocean Lab, a sea-life charity with a waterside cafe, shop and centre. It is very eco there and there is an aquarium by the beach. The Sea Trust controls half of it and there are open days, beach clean-ups and a gift shop with eco-goodies. One of my worries is that it will close down because they're not earning enough money. All the local shops are closing, everyone is buying online. I am very worried.

My friend had a good idea that you should plant a tree every special day of your life like birthdays, Easter, Halloween, Christmas; if you don't have a garden you can put up an indoor plant. It is a very good idea and I think that there should be a day every summer holiday when everyone plants as many plants and trees as they can. Also, there should be a day when everyone does litter picking.

I was worried, especially when I found out about people taking fossil fuels from the ground and I know not to buy so much now. I was very scared but then I noticed there was HOPE.

Everyone is solving things, everyone is realising what is happening but I believe to save our planet we need to work together and not be an individual.

I would like to see much more plastic recycling. We need more solar panels and plant pots in yards and gardens with trees and flowers instead of concrete. We need a non-electric day!

I have realised that in your speeches you are saying the same thing to people over and over, but it seems as if none of the politicians who are in charge of big decisions are listening. I think they are the ones who need to change.

I see things in a different way so if someone says, 'It will not make a difference, it's just one bottle', it would seem to me like all they are saying is let's throw a bottle on the ground and leave it there so someone else can deal with it.

I've given up on being polite, I say what I think and when it comes to saving the planet me and my friends are planning on action. One day I would like to invent an eco-robot that picks up litter. The robot would have a long tongue that comes out and picks up the rubbish. It would have four wheels instead of feet and soap and water inside it to wash the rubbish. It would be solar powered.

These are my ideas and worries. Thank you so much for listening to me. Keep on fighting,

From Beth x

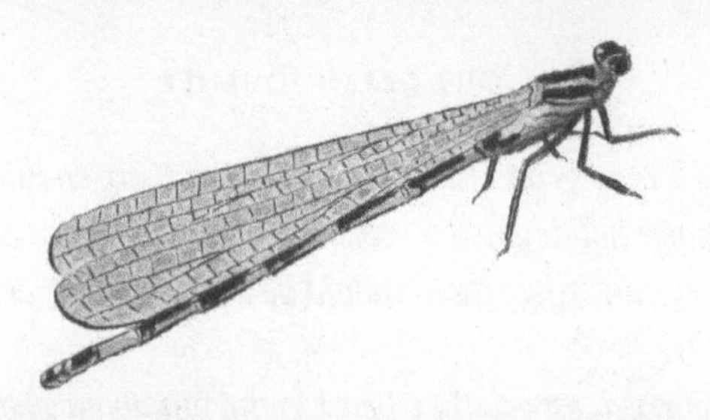

BETHAN'S SUSTAINABLE SCHOOLS MANIFESTO

As a schoolgirl I have my opinions on eco schools, and I feel my opinions and ideas count so here are some of them. They can be adapted for a workplace, home or college.

- Schools should grow their own foods: potatoes, leeks, tomatoes, etc.
- Schools should do local workshops to raise awareness of world problems and fundraise for solar panels or eco school supplies. (At my school, Ysgol wdig / Goodwick Community School, this went very well and we even had some local eco heroes coming in to talk to us.)
- Schools should do litter-picks in the school grounds and public areas.
- The younger kids in our school definitely enjoyed making animal habitats such as bird homes, insect hotels, hoverfly pools, etc.
- Being a vegetarian or vegan definitely makes a big difference so why not suggest a vegetarian day every week or, even better, become a vegetarian.
- In our school we have a small wild area we could expand and I think all schools should start a small one.

- In our school it is hard to give up single-use plastics but we can give it a shot.
- I think it would make a difference to do eco assemblies to inform everyone about what you are doing as a school; you could even have some visitors coming in to talk.
- A good idea I got from an Eco Schools conference is to write a letter to your local MP; this is a great classroom activity.
- In 2019, my best friends, Eliza and Leila, and I started a litter-picking club with the school litter-pickers. It would make a big difference if you made one too.
- If you live close to your school why not cycle or walk to school; if you live far away then catch a bus.
- It is very important that school kids are well informed about world problems. What better way to do it than to go on eco-school trips to places such as local companies, farms and beaches.
- On warm days, instead of being inside why not go outside for outdoor classes?
- Starting a mini farm is such a good idea; you could have vegetables and chickens – the only thing is someone has to look after them during the holidays.
- Plastic-free lunches and packed lunches are an awesome idea (I got this idea from an Eco-schools conference).
- Another good idea is to go to school with a reusable water bottle as then the school won't need to buy plastic cups.
- Our school makes art out of litter instead of using new materials.
- Encourage the school to use fewer chemicals by using eco cleaning products.
- In my garden at home we have a compost bin. All schools produce so much green waste they should compost it.
- Books swaps, clothes swaps and toys swaps at school would be very worthwhile. We do this at our school.

RECOMMENDED READING

A Life on Our Planet: My Witness Statement and a Vision for the Future by David Attenborough (Ebury Press, 2020)

A Wild Child's Guide to Endangered Animals by Millie Marotta (Particular Books, 2019)

Climate Rebels by Ben Lerwill (Penguin, 2021)

How to Give Up Plastic by Will McCallum (Penguin Life, 2019)

How to Save the World for Free by Natalie Fee (Laurence, 2019)

Love Earth Now: The Power of Doing One Thing Every Day by Cheryl Leutjen (Mango, 2018)

No One Is Too Small to Make a Difference by Greta Thunberg (Penguin, 2019)

No.More.Plastic by Martin Dorey (Ebury, 2018)

Six Weeks to Zero Waste: A Simple Plan for Life by Kate Arnell (Gaia, 2020)

The Hidden Life of Trees by Peter Wohlleben (William Collins, 2017)

The Sixth Extinction: An Unnatural History by Elizabeth Kolbert (Bloomsbury, 2015)

The Joyful Environmentalist: How to Practise without Preaching by Isabel Losada (Watkins, 2020)

The Sustainable(ish) Living Guide by Jen Gale (Green Tree, 2020)

There is No Planet B: A Handbook for the Make or Break Years by Mike Berners-Lee (Cambridge University Press, 2019)

This Changes Everything: Capitalism vs. The Climate by Naomi Klein (Penguin, 2015)

This is Not a Drill: An Extinction Rebellion Handbook by Extinction Rebellion (Penguin, 2019)

ILLUSTRATED BOOKS FOR CHILDREN:

Duffy's Lucky Escape by Ellie Jackson, illustrated by Liz Oldmeadow (Ellie Jackson, 2017)

Little Turtle Turns the Tide by Lauren Davies, illustrated by Nico Williams (Orca Publications Ltd, 2019)

Marli's Tangled Tale by Ellie Jackson, illustrated by Laura Callwood (Ellie Jackson, 2017)

Nelson's Dangerous Dive by Ellie Jackson, illustrated by Laura Callwood (Ellie Jackson, 2018)

Poseidon's Plastic Problem by Ellie Boyle, illustrated by Lizzie Nightingale (2020)

AN A–Z OF ENVIRONMENTAL ORGANISATIONS

If you intend to write, please enclose a stamped addressed envelope (recycled, of course) to make sure you get a reply. Many groups are volunteer-based and will need your support.

Action Against Allergy
'We are a national charity dedicated to providing help and support to anyone with an allergic medical condition.'
Address: 7 Strawberry Hill Road, Twickenham, TW1 4QB, UK
Phone: (+44) (0) 20 8892 4949
Email: help@actionagainstallergy.org
Website: www.actionagainstallergy.org

Action Against Desertification
'Action Against Desertification is an initiative to tackle the detrimental social, economic and environmental impact of land degradation and desertification.'
Address: Viale delle Terme di Caracalla, 00153 Rome, Italy
Phone: (+39) 06 57051
Email: action-against-desertification@fao.org
Website: www.fao.org

ActionAid
'ActionAid is an international charity that works with women and girls living in poverty.'
Address: 33–39 Bowling Green Lane, Farringdon,

London, EC1R 0BJ, UK
Phone: (+44) (0) 1460 238000
Email: supportercontact@actionaid.org
Website: www.actionaid.org.uk

Action for Insects (The Wildlife Trusts)
'The Wildlife Trusts is a grassroots movement of people from a wide range of backgrounds and all walks of life, who believe that we need nature and nature needs us.'
Address: The Kiln, Mather Road, Newark, NG24 1WT, UK
Phone: (+44) (0) 1636 677711
Email: enquiry@wildlifetrusts.org
Website: www.wildlifetrusts.org

Action For Rail
'Its aim is to work with passenger groups, rail campaigners and environmentalists to campaign against cuts to rail services and staffing and to promote the case for integrated, national rail under public ownership.'
Website: www.actionforrail.org

Action on Smoking and Health (ASH)
'Action on Smoking and Health (ASH) is a public health charity that works to eliminate the harm caused by tobacco. We do not attack smokers or condemn smoking.'
Address: 6th Floor, Suites 59–63, New House,
67–68 Hatton Garden, London, EC1N 8JY, UK
Phone: (+44) (0) 20 7404 0242
Email: enquiries@ash.org.uk
Website: www.ash.org.uk

Allergy UK
'We are the leading national patient charity for people living with all types of allergy. Our vision is to help improve the lives of the millions of people with allergic disease.'
Address: Planwell House, LEFA Business Park,
Edgington Way, Sidcup, Kent, DA14 5BH, UK
Phone: (+44) (0) 1322 619898

Email: info@allergyuk.org
Website: www.allergyuk.org

Anaphylaxis Campaign
'Our ultimate aim is to create a safe environment for all people with allergies.'
Address: 1 Alexandra Road, Farnborough, Hampshire, GU14 6BU, UK
Phone: (+44) (0) 1252 546100
Email: admin@anaphylaxis.org.uk
Website: www.anaphylaxis.org.uk

Animal Aid
'Animal Aid was founded in January 1977 to work, by all peaceful means, for an end to animal cruelty.'
Address: The Old Chapel, Bradford Street, Tonbridge, Kent, TN9 1AW, UK
Phone: (+44) (0) 1732 364546
Email: info@animalaid.org.uk
Website: www.animalaid.org.uk

Animal Free Research UK
'To create a world where human diseases are cured faster without animal suffering.'
Address: Phoenix Yard, 65 King's Cross Road, London, WC1X 9LW, UK
Phone: (+44) (0) 20 8054 9700
Email: hello@animalfreeresearchuk.org
Website: www.animalfreeresearchuk.org

Animal Outlook
'We're exposing the truth, delivering justice, revolutionising food systems, and empowering others to stand up for animals by leaving them off our plates.'
Address: P.O. Box 9773, Washington, DC 20016, USA
Phone: (+1) 301 891 2458
Email: info@animaloutlook.org
Website: www.animaloutlook.org

Banana Link
'Banana Link campaigns for fair and equitable production and trade in bananas and pineapples based on environmental, social and economic sustainability.'
Address: 42–58 St George's Street, Norwich, NR3 1AB, UK
Phone: (+44) (0) 1603 765670
Website: www.bananalink.org.uk

Bikeability – National Cycling Proficiency Test
'Bikeability is today's cycle training programme. It's about gaining practical skills and understanding how to cycle on today's roads.'
Address: The Bikeability Trust, ideaSpace City, 3 Laundress Lane, Cambridge, CB2 1SD, UK
Phone: (+44) (0) 800 8491017
Email: contact@bikeabilitytrust.org
Website: www.bikeability.org.uk

Bioregional Group – One Planet Living
'We provide a range of sustainability consultancy services that make it easier for people to live sustainable lives within the planet's boundaries.'
Address: Bioregional, BedZED Centre, 24 Helios Road, Wallington, London, SM6 7BZ, UK
Phone: (+44) (0) 20 8404 4880
Email: info@bioregional.com
Website: www.bioregional.com

Black Environment Network
'BEN works to enable full ethnic participation in the built and natural environment.'
Address: The Warehouse, 54–57 Allison Street, Digbeth, Birmingham, B5 5TH, UK
Phone: (+44) (0) 121 6436387
Email: ukoffice@ben-network.org.uk
Website: www.ben-network.org.uk

Breast Cancer UK
'We give practical advice on how people can reduce their risk of getting breast cancer through making simple changes in their lives.'
Address: St George's House, 14 George Street, Huntingdon, Cambridgeshire, PE29 3GH, UK
Phone: (+44) (0) 845 680 1322
Email: enquiries@breastcanceruk.org.uk
Website: www.breastcanceruk.org.uk

Brilliant Earth
'We are passionate about cultivating a more transparent, sustainable, and compassionate jewellery industry.'
Website: www.brilliantearth.com

British Ecological Society
'We are the British Ecological Society: the oldest ecological society in the world.'
Address: 42 Wharf Road, London, N1 7GS, UK
Phone: (+44) (0) 20 3994 8282
Email: hello@britishecologicalsociety.org
Website: www.britishecologicalsociety.org

British Lung Foundation
'With your support, we'll make sure that one day everyone breathes clean air with healthy lungs.'
Address: 73–75 Goswell Road, EC1V 7ER, London, UK
Phone: (+44) (0) 3000 030 555
Website: www.blf.org.uk

The British Nutrition Foundation
'We provide impartial, evidence-based information on food and nutrition. Our core purpose is to make nutrition science accessible to all.'
Address: New Derwent House, 69–73 Theobalds Road, London, WC1X 8TA, UK
Phone: (+44) (0) 20 7557 7930
Email: postbox@nutrition.org.uk
Website: www.nutrition.org.uk

British Plastics Federation (BPF)
'The British Plastics Federation is the world's longest running plastics trade association, established in 1933 to represent the UK industry.'
Address: 5–6 Bath Place, Rivington Street, Hackney, London, EC2A 3JE, UK
Phone: (+44) (0) 20 7457 5000
Email: reception@bpf.co.uk
Website: www.bpf.co.uk

Buglife
'Buglife is the only organisation in Europe devoted to the conservation of all invertebrates.'
Address: The Lindens, 86 Lincoln Road, Peterborough, PE1 2SN, UK
Phone: (+44) (0) 1733 201 210
Email: info@buglife.org.uk
Website: www.buglife.org.uk

Bus Users
'We're making sure that everyone has access to the best possible transport links.'
Address: Victoria Charity Centre, 11 Belgrave Road, London, SW1V 1RB, UK
Phone: (+44) (0) 300 111 0001
Email: enquiries@bususers.org
Website: www.bususers.org

CAFOD
'We are part of one of the largest aid networks in the world. Because of our global reach and local presence, together we have the potential to reach everyone.'
Address: Romero House, 55 Westminster Bridge Road, London, SE1 7JB, UK
Phone: (+44) (0) 303 303 3030
Email: cafod@cafod.org.uk
Website: www.cafod.org.uk

Campaign for Better Transport
'Campaign for Better Transport's vision is for all communities to have access to high quality, sustainable transport that meets their needs, improves quality of life and protects the environment.'
Address: 70 Cowcross Street, London, EC1M 6EJ, UK
Phone: (+44) (0) 300 303 3824
Email: communications@bettertransport.org.uk
Website: www.bettertransport.org.uk

Campaign for Nuclear Disarmament (CND)
'CND campaigns non-violently to achieve British nuclear disarmament – to get rid of the Trident nuclear weapons system and stop its replacement.'
Address: Mordechai Vanunu House, 162 Holloway Road, London, N7 8DQ, UK
Phone: (+44) (0) 20 7700 2393
Email: enquiries@cnduk.org
Website: www.cnduk.org

The Campaign for Safe Cosmetics
'The Campaign has educated millions of people about the problem of toxic chemicals in cosmetics, which has led to an increased demand for safer products in the marketplace.'
Address: 1388 Sutter Street, Suite 400, San Francisco, CA 94109-5400, USA
Email: safecosmetics@bcpp.org
Website: www.safecosmetics.org

Campaign to Protect Rural England (CPRE)
'We're working for a countryside that's rich in nature, accessible to everyone and playing a crucial role in responding to the climate emergency.'
Address: 5–11 Lavington Street, London, SE1 0NZ, UK
Phone: (+44) (0) 20 7981 2800
Email: info@cpre.org.uk
Website: www.cpre.org.uk

Carbon Footprint Ltd
'A breath of fresh air for carbon & sustainability.'
Address: Belvedere House, Basing View, Basingstoke, Hampshire, RG21 4HG, UK
Phone: (+44) (0) 1256 592599
Email: info@carbonfootprint.com
Website: www.carbonfootprint.com

Centre for Alternative Technology (CAT)
'CAT is an educational charity dedicated to researching and communicating positive solutions for environmental change.'
Address: Llwyngwern Quarry, Pantperthog, Machynlleth, SY20 9AZ, UK
Phone: (+44) (0) 1654 705950
Email: contactus@cat.org.uk
Website: www.cat.org.uk

Centres for Cities
'To promote education for the public benefit in issues of economics and public policy in relation to cities and towns in the United Kingdom and elsewhere in the world and to promote for the public benefit research in the aforementioned fields and to publish the useful results of such research.'
Address: 2nd Floor, 9 Holyrood, London, SE1 2EL, UK
Phone: (+44) (0) 20 7803 4300
Email: info@centreforcities.org
Website: www.centreforcities.org

The Children's Society
'We listen. We support. We act. Because we believe no child should feel alone.'
Address: Whitecross Studios, 50 Banner Street, London, EC1Y 8ST, UK
Phone: (+44) (0) 300 303 7000
Email: supportercare@childrenssociety.org.uk
Website: www.childrenssociety.org.uk

City to Sea
'An award-winning, environmental not-for-profit on a mission to stop plastic pollution at source. We run people-powered, community-serving campaigns that help individuals and businesses to change the world whilst having some fun on the way.'
Address: Albion Dockside Building, Hanover Place, Bristol BS1 6UT
Phone: (+44) (0) 117 422 4522
Email: admin@citytosea.org.uk
Website: www.citytosea.org.uk and www.refill.org.uk

Clean Clothes Campaign
'Clean Clothes Campaign is a global network dedicated to improving working conditions and empowering workers in the global garment and sportswear industries.'
Address: Postbus 11584, 1001 GN Amsterdam, the Netherlands
Email: info@cleanclothes.org
Website: www.cleanclothes.org

ClientEarth
'ClientEarth is a charity that uses the power of the law to protect the planet and the people who live on it.'
Address: Fieldworks, 274 Richmond Road, Hackney, London, E8 3QW, UK
Phone: (+44) (0) 20 7749 5970
Email: info@clientearth.org
Website: www.clientearth.org

Climate Action Network (International) (CAN)
'The Climate Action Network is a worldwide network of over 1300 Non-Governmental Organizations (NGOs) in more than 120 countries, working to promote government and individual action to limit human-induced climate change to ecologically sustainable levels.'
Address: Kaiserstraße 201, Bonn, 53113, Germany
Email: administration@climatenetwork.org
Website: www.climatenetwork.org

Climate and Migration Coalition
'Protecting people who move due to the impacts of climate change.'
Address: The Old Music Hall, 106–108 Cowley Road, Cowley, Oxford, OX4 1JE, UK
Phone: (+44) (0) 1865 403334
Website: www.climatemigration.org.uk

Climate Coalition
'A cleaner, greener future is within reach. We are the UK's largest group of people dedicated to action against climate change.'
Address: Romero House, 55 Westminster Bridge Road, London, SE1 7JB, UK
Phone: (+44) (0) 20 7870 2213
Website: www.theclimatecoalition.org

Climate Psychology Alliance
'A group for anyone interested in making connections between in-depth psychology and climate change.'
Email: contact@climatepsychologyalliance.org
Website: www.climatepsychologyalliance.org

The Coalition for Consumer Information on Cosmetics – Leaping Bunny Programme
'We work with companies to help make shopping for animal-friendly products easier and more trustworthy.'
Address: 801 Old York Road, Suite 204, Jenkintown, PA 19046, USA
Email: info@leapingbunny.org
Website: www.leapingbunny.org

Common Ground
'Common Ground was founded in 1983 for people to seek imaginative ways to engage people with their local environment.'
Address: Lower Dairy, Toller Fratrum, Dorset, DT2 0EL, UK
Phone: (+44) (0) 1300 321536
Website: www.commonground.org.uk

Compassion in World Farming
'Today we campaign peacefully to end all factory farming practices.'
Address: River Court, Mill Lane, Godalming, Surrey, GU7 1EZ, UK
Phone: (+44) (0) 1483 521953
Website: www.ciwf.org.uk

Conservation International – Footprint Calculator
'Through science, policy and partnerships, Conservation International is protecting perhaps humanity's biggest ally in the fight against climate change: nature.'
Address: 2011 Crystal Drive, Suite 600, Arlington, VA 22202, USA
Phone: (+1) 703 341 2400
Email: community@conservation.org
Website: www.conservation.org/carbon-footprint-calculator

The Conservation Volunteers
'TCV works together with communities to deliver practical solutions to the real-life challenges they face.'
Address: Sedum House, Mallard Way, Doncaster, DN4 8DB, UK
Phone: (+44) (0) 1302 388 883
Email: information@tcv.org.uk
Website: www.tcv.org.uk

Consumer Council for Water
'Our team is vastly experienced in providing consumers with free advice and support on every aspect of their water and sewerage services.'
Address: 1st Floor, Victoria Square House, Victoria Square, Birmingham, B2 4AJ, UK
Phone: (+44) (0) 300 034 2222
Email: pr@ccwater.org.uk
Website: www.ccwater.org.uk

Consumer Goods Forum
'Positive business actions that benefit both people and the planet.'
Address: 47–53 rue Raspail, 92300, Levallois-Perret, France
Phone: (+33) 1 82 00 95 95
Email: info@theconsumergoodsforum.com
Website: www.theconsumergoodsforum.com

The Convention on International Trade in Endangered Species of Wild Fauna and Flora (CITES)
'Its aim is to ensure that international trade in specimens of wild animals and plants does not threaten their survival.'
Address: CITES Secretariat, Palais des Nations, Avenue de la Paix 8–14, 1211 Genève 10, Switzerland
Phone: (+41) (0) 22 917 81 39/40
Email: info@cites.org
Website: www.cites.org

Cornwall Wildlife Trust
'We want to create safe havens in which wildlife can thrive and from which it can spread.'
Address: Five Acres, Allet, Truro, Cornwall, TR4 9DJ, UK
Phone: (+44) (0) 1872 273 939
Email: info@cornwallwildlifetrust.org.uk
Website: www.cornwallwildlifetrust.org.uk

Cotton Campaign
'The Cotton Campaign has advocated with governments, companies and investors to use their leverage to end this continuous and systematic human rights violation.'
Address: 1634 I Street NW, Suite 1000, Washington, DC 20006, USA
Phone: (+1) 202 347 4100
Email: cottoncampaigncoordinator@gmail.com
Website: www.cottoncampaign.org

Counting Animals
'Counting Animals is a blog that aims to collect, generate, analyse, organize, present and interpret quantitative information related to the animal advocacy movement.'
Address: 199 Midfield Road, Ardmore, PA 19003-3212, USA
Email: harish.sethu@gmail.com
Website: www.countinganimals.com

Critical Mass
'A Critical Mass is intended to attract an ever growing number of cycle riders, who through strength of numbers are visible and assert their right to be there.' Last Friday of every month in cities worldwide.
Website: www.critical-mass.org

Cruelty Free International Trust
'Cruelty Free International is the leading organisation working to create a world where nobody wants or believes we need to experiment on animals.'
Address: 16a Crane Grove, London, N7 8NN, UK
Phone: (+44) (0) 20 7700 4888
Email: info@crueltyfreeinternational.org
Website: www.crueltyfreeinternational.org

Cycling UK
'Cycling UK's vision is of a healthier, happier and cleaner world, because more people cycle.'
Address: Parklands, Railton Road, Guildford, Surrey, GU2 9JX, UK
Phone: (+44) (0) 1483 238300
Email: cycling@cyclinguk.org
Website: www.cyclinguk.org

Department for Environment, Food & Rural Affairs
'The Department is responsible for matters including environment protection and conservation of biodiversity as well as energy policy.'

Address: Nobel House, 17 Smith Square, Westminster, London, SW1P 3RJ, UK
Phone: (+44) (0) 345 933 5577
Email: defra@public.govdelivery.com
Website: www.gov.uk/government/organisations/department-for-environment-food-rural-affairs

Department for Transport
'We work with our agencies and partners to support the transport network that helps the UK's businesses and gets people and goods travelling around the country.'
Address: Great Minster House, 33 Horseferry Road, Westminster, London, SW1P 4DR, UK
Phone: (+44) (0) 300 330 3000
Email: contactdft@dft.gov.uk
Website: www.gov.uk/government/organisations/department-for-transport

Dogs Trust
'Our mission is to bring about the day when all dogs can enjoy a happy life, free from the threat of unnecessary destruction.'
Address: 17 Wakley Street, London, EC1V 7RQ, UK
Phone: (+44) (0) 20 7837 0006
Email: contactcentre@dogstrust.org.uk
Website: www.dogstrust.org.uk

Eco-Age
'We are committed to helping brands to create a culture of purpose & sustainability.'
Address: Gemini House, 334–336 King Street, London, W6 0RR, UK
Phone: (+44) (0) 20 8995 7611
Email: info@eco-age.com
Website: www.eco-age.com

The Ellen MacArthur Foundation
'The Ellen MacArthur Foundation works to inspire a generation to re-think, re-design and build a positive future circular economy.'

Address: The Sail Loft, 42 Medina Road, Cowes, Isle of Wight, PO31 7BX, UK
Phone: (+44) (0) 1983 296 463
Email: info@ellenmacarthurfoundation.org
Website: www.ellenmacarthurfoundation.org

Energy Saving Trust
'We are a leading and trusted organisation helping people save energy every day.'
Address: 30 North Colonnade, Canary Wharf, London, E14 5GP, UK
Phone: (+44) (0) 800 444 202
Website: www.energysavingtrust.org.uk

Enjoy The Countryside
'Our aim is to raise awareness of the dangers in the British countryside so that everyone can enjoy it to the maximum, whilst staying safe.'
Website: www.enjoythecountryside.com

Environment and Conservation Volunteering
'Our mission is to change the face of volunteer travel.'
Address: International Volunteer HQ, P.O. Box 8273, New Plymouth 4340, New Zealand
Email: info@volunteerhq.org
Website: www.volunteerhq.org

Environmental Investigation Agency
'We investigate and campaign against environmental crime and abuse.'
Address: 62–63 Upper Street, The Angel, London, N1 0NY, UK
Phone: (+44) (0) 20 7354 7960
Email: supporterservices@eia-international.org
Website: www.eia-international.org

Environmental Law Association
'ELA is the UK forum which aims to make better law for the environment and to improve understanding and awareness of environmental law.'
Address: One Glass Wharf, Bristol, BS2 0ZX, UK
Website: www.ukela.org

Fairtrade Foundation
'Fairtrade is about better prices, decent working conditions, local sustainability, and fair terms of trade for farmers and workers in the developing world.'
Address: 5.7 The Loom, 14 Gowers Walk, London, E1 8PY, UK
Phone: (+44) (0) 20 7405 5942
Email: mail@fairtrade.org.uk
Website: www.fairtrade.org.uk

FareShare
'FareShare is the UK's largest charity fighting hunger and food waste. We save good food from going to waste and redistribute it to frontline charities.'
Address: Unit 7, Deptford Trading Estate, Blackhorse Road, London, SE8 5HY, UK
Phone: (+44) (0) 131 608 0967
Email: enquiries@fareshare.org.uk
Website: www.fareshare.org.uk

Fashion for Good
'We believe that Good Fashion is not only possible, it is within reach.'
Email: connect@fashionforgood.com
Website: www.fashionforgood.com

Fashion Revolution
'A global fashion industry that conserves and restores the environment and values people over growth and profit.'
Address: 70 Derby Street, Leek, Staffordshire ST13 5AJ, UK
Website: www.fashionrevolution.org

Feed Your Happy

'Feed Your Happy campaign invites you to add happiness to the world and your life by feeding your happy with organic.'
Address: 18 Rooksbury Road, Andover, Hampshire, SP10 2LW, UK
Email: info@organictradeboard.co.uk
Website: www.feedyourhappy.co.uk

The Food Commission

'The Food Commission (UK) Ltd is a not-for-profit company which campaigns for healthier, safer, sustainable food in the UK.'
Address: 94 White Lion Street, London, N1 9PF, UK
Phone: (+44) (0) 20 7837 2250
Email: info@foodcomm.co.uk
Website: www.foodcomm.org.uk

Food Cycle

'We support people who are hungry and lonely by serving tasty lunches and dinners every single day in towns and cities across the country.'
Address: New Covent Garden Market, FoodCycle, 2.16 The Food Exchange, London, SW8 5EL, UK
Phone: (+44) (0) 20 7729 2775
Email: hello@foodcycle.org.uk
Website: www.foodcycle.org.uk

Food Empowerment Project

'Food Empowerment Project seeks to create a more just and sustainable world by recognizing the power of one's food choices.'
Address: P.O. Box 7322, Cotati, CA 94931, USA
Phone: (+1) 707 779 8004
Email: info@foodispower.org
Website: www.foodispower.org

Food for Life

'Food for Life brings schools, nurseries, hospitals and care homes, and their surrounding communities together around the core ethos of healthy, tasty and sustainable food.'
Address: Soil Association, Food for Life, Spear House, 51 Victoria Street, Bristol, BS1 6AD, UK
Phone: (+44) (0) 117 314 5180
Email: ffl@foodforlife.org.uk
Website: www.foodforlife.org.uk

FoodSave

'FoodSave is a project to help small and medium-sized food businesses in London reduce their food waste.'
Phone: (+44) (0) 20 7479 4244
Email: hello@thesra.org
Website: www.foodsave.org

Food Standards Agency

'Our job is to use our expertise and influence so that people can trust that the food they buy and eat is safe and what it says it is.'
Address: Floors 6 and 7, Clive House, 70 Petty France, London, SW1H 9EX, UK
Phone: (+44) (0) 330 332 7149
Email: helpline@food.gov.uk
Website: www.food.gov.uk

Food Tank

'Food Tank is focused on building a global community for safe, healthy, nourished eaters.'
Address: 1915 Bank Street, Baltimore, MD 21231, USA
Email: danielle@foodtank.com
Website: www.foodtank.com

Frank Water

'Our small organisation aims to address a big problem, providing safe drinking water to the 663 million people worldwide who still lack access to this basic human right.'
Address: 11 Elmdale Road, Bristol, BS8 1SL, UK

Phone: (+44) (0) 117 329 4846
Email: hello@frankwater.com
Website: www.frankwater.com

Freedom for Animals
'We take action to end the captivity of animals, especially those used for entertaining the public in zoos, circuses and the media industry.'
Address: P.O. Box 591, Manchester, M12 0DP, UK
Phone: (+44) (0) 845 330 3911
Email: office@freedomforanimals.org.uk
Website: www.freedomforanimals.org.uk

freedom4girls UK
'We actively support women and girls in both the UK and in East Africa, who struggle to access safe menstrual hygiene protection.'
Address: 15 Roundhay View, Leeds, LS8 4DX, UK
Email: info@freedom4girls.co.uk
Website: www.freedom4girls.co.uk

Fridays For Future
'A global movement that began in August 2018, when 15-year-old Greta Thunberg and other young activists began a school strike for climate.'
Website: www.fridaysforfuture.org

Friends of the Earth England, Wales and Northern Ireland
'We are part of an international community dedicated to protecting the natural world and the wellbeing of everyone in it.'
Address: The Printworks, 139 Clapham Road, London, SW9 0HP, UK
Phone: (+44) (0) 20 7490 1555
Email: info@foe.co.uk
Website: www.friendsoftheearth.uk

Fruit Factory – Brighton Permaculture Trust
'Brighton Permaculture Trust is a charity that promotes greener lifestyles and sustainable development through design.'
Address: The Fruit Factory, Stanmer Village, Stanmer, Brighton, BN1 9PZ, UK
Phone: (+44) (0) 7746 185 927
Email: admin@brightonpermaculture.org.uk
Website: www.brightonpermaculture.org.uk

GiveMeTap
'Our mission is to make clean drinking water available to everyone, everywhere.'
Address: 12 Clarendon Road, Studio B, London, N18 2AJ, UK
Website: www.givemetap.com

Global Fashion Agenda
'Global Fashion Agenda is the foremost leadership forum for industry collaboration on fashion sustainability.'
Address: Frederiksholms Kanal 30-A6, DK-1220, Copenhagen, Denmark
Phone: (+45) 70 20 30 68
Email: info@globalfashionagenda.com
Website: www.globalfashionagenda.com

Global Footprint Network
'We enable our vision through our mission: to help end ecological overshoot by making ecological limits central to decision-making.'
Address: 1528 Webster Street, Suite 11, Oakland, CA 94612, USA
Phone: (+1) 510 839 8879
Website: www.footprintnetwork.org

Good Goddess Project
'The Good Goddess project is about touching hearts with Love, planting seeds of hope for all women. Our work includes taking love to where there is none; women's refuges, refugee women, food banks, providing eco-friendly sanitary ware and empowering women's circles. Together we are the Change.'

Address: c/o 20 Broadway, Fulwood, Preston, PR2 9TH, UK
Facebook: The Good Goddess Project
Email: thegoodgoddessproject@gmail.com

Good On You
'We are a group of campaigners, fashion professionals, scientists, writers and developers who have come together to drive environmental change.'
Online campaign.
Email: info@goodonyou.eco
Website: www.goodonyou.eco

The Green Age
'TheGreenAge is the UK's premier energy saving advice portal, covering heating, insulation and renewable technologies.'
Address: Unit 1, Kingston Business Centre, Fullers Way, South Chessington, Surrey, KT9 1DQ, UK
Phone: (+44) (0) 20 8144 0897
Email: mailbox@thegreenage.co.uk
Website: www.thegreenage.co.uk

Green Energy UK
'We are not like other green energy providers. We challenge the idea that 'being green' is all about making a big commitment, about making an extra effort, and paying extra costs.'
Address: Black Swan House, 23 Baldock Street, Ware, Hertfordshire, SG12 9DH, UK
Phone: (+44) (0) 1920 486156
Email: help@greenenergyuk.com
Website: www.greenenergyuk.com

Greener Journeys
'Greener Journeys is dedicated to encouraging more sustainable travel.'
Address: 27 Beaufort Court, Admirals Way, London, E14 9XL, UK
Email: contact@greenerjourneys.com
Website: www.greenerjourneys.com

Green Futures Festivals Radio/Kingston Green Radio
'A vision to give a voice to the Kingston community while highlighting the urgent need for environmental awareness locally and around the world.'
Address: Kingston Environment Centre, 1 Kingston Road, New Malden, Surrey, KT3 3PE, UK
Phone: (+44) (0) 20 8942 5251
Email: studio@www.gffradio.org.uk
Website: www.greenfuturesfestivals.org.uk

Greenpeace UK
'Greenpeace is a movement of people who are passionate about defending the natural world from destruction.'
Address: Canonbury Villas, London, N1 2PN, UK
Phone: (+44) (0) 20 7865 8100
Email: press.uk@greenpeace.org
Website: www.greenpeace.org.uk

Healthy Air Campaign
'We are calling for urgent action on air pollution.'
Address: Healthy Air Campaign, ClientEarth, 274 Richmond Road, London, E8 3QW, UK
Phone: (+44) (0) 207 749 5979
Email: alee@clientearth.org
Website: www.healthyair.org.uk

Humane Society International
'Humane Society International works around the globe to promote the human–animal bond, protect street animals, support farm animal welfare, stop wildlife abuse, eliminate painful animal testing, respond to natural disasters and confront cruelty to animals in all of its forms.'
Address: 5 Underwood Street, Hoxton, London, N1 7LY, UK
Phone: (+44) (0) 20 7490 5288
Email: info@hsiuk.org
Website: www.hsi.org

Human Rights Watch
'Human Rights Watch investigates and reports on abuses happening in all corners of the world'
Address: Audrey House, 16–20 Ely Place, London, EC1N 6SN, UK
Phone: (+44) (0) 20 7618 4700
Website: www.hrw.org

Independent Food Aid Network
'The Independent Food Aid Network is a network of independent, grassroots food aid providers working together to secure food security for all.'
Address: 58 Standen Road, London, SW18 5TQ, UK
Phone: (+44) (0) 20 8704 4557
Email: ifan.uk@gmail.com
Website: www.foodaidnetwork.org.uk

The Intergovernmental Panel on Climate Change
'The IPCC provides regular assessments of the scientific basis of climate change, its impacts and future risks, and options for adaptation and mitigation.'
Website: www.ipcc.ch

International Volunteer HQ
'Our mission is to change the face of volunteer travel.'
Phone: (+44) (0) 800 031 8376
Email info@volunteerhq.org
Website: www.volunteerhq.org

Karma
'The food industry is unsustainable in its attitudes towards waste. Karma is changing this whilst helping to turn surplus into a business opportunity.'
Address: Sweden House, 5 Upper Montagu Street, London, W1H 2AG, UK
Email: support@karma.life
Website: www.karma.life

Kaya Responsible Travel
'Kaya's mission is to promote sustainable social, environmental and economic development, empower communities and cultivate educated, compassionate global citizens through responsible travel.'
Address: The Arches, North Campus, Sackville Street, Manchester, M60 1QD, UK
Phone: (+44) (0) 161 870 6212
Email: info@kayavolunteer.com
Website: www.kayavolunteer.com

Keep Britain Tidy
'Keep Britain Tidy is an independent charity with three goals – to eliminate litter, end waste and improve places.'
Address: Elizabeth House, The Pier, Wigan, WN3 4EX, UK
Phone: (+44) (0) 1942 612 621
Email: news@keepbritaintidy.org
Website: www.keepbritaintidy.org

Kent Wildlife Trust
'Together with our members, supporters, volunteers and partners, we work to protect wildlife across the county.'
Address: Tyland Barn, Chatham Road, Sandling, Maidstone, ME14 3BD, UK
Phone: (+44) (0) 1622 662012
Email: membership@kentwildlife.org.uk
Website: www.kentwildlifetrust.org.uk

Liberty Human Rights
'We challenge injustice, defend freedom and campaign to make sure everyone in the UK is treated fairly.'
Address: Liberty House, 26–30 Strutton Ground, London, SW1P 2HR, UK
Phone: (+44) (0) 20 7403 3888
Email: comms@libertyhumanrights.org.uk
Website: www.libertyhumanrights.org.uk

Living Streets
'We want a nation where walking is the natural choice for everyday local journeys.'
Address: America House, 2 America Square, London, EC3N 2LU, UK
Phone: (+44) (0) 20 7377 4900
Email: info@livingstreets.org.uk
Website: www.livingstreets.org.uk

London Cycling Campaign
'We strive for a city that encourages Londoners to cycle, creating a healthier and happier place for everyone.'
Address: Unit 201, Metropolitan Wharf, 70 Wapping Wall, London, E1W 3SS, UK
Phone: (+44) (0) 20 7234 9310
Email: info@lcc.org.uk
Website: www.lcc.org.uk

London Hazards Centre
'London Hazards Centre works with trade unions, tenants groups and community based campaigns.'
Address: 227 Seven Sisters Road, Finsbury Park, London, N4 2DA, UK
Phone: (+44) (0) 20 7527 5107
Email: mail@lhc.org.uk
Website: www.lhc.org.uk

Love Food Hate Waste
'Love Food Hate Waste aims to raise awareness of the need to reduce food waste and help us take action.'
Address: Second Floor, Blenheim Court, 19 George Street, Banbury, Oxfordshire, OX16 5BH, UK
Website: www.lovefoodhatewaste.com

Love Water Campaign
'Love Water is a major campaign involving more than 40 environmental groups, charities, water companies and regulators, aimed at getting the British public involved in safeguarding water resources for future generations.'

Address: Water UK, 3rd Floor, 36 Broadway, Westminster, London, SW1H 0BH, UK
Phone: (+44) (0)20 7344 1844
Email: webadmin@water.org.uk
Website: www.water.org.uk

Marine Conservation Society
'The Marine Conservation Society is the UK's leading charity for the protection of our seas, shores and wildlife.'
Address: Overross House, Ross Park, Ross-on-Wye, Herefordshire, HR9 7US, UK
Phone: (+44) (0) 1989 566017
Email: info@mcsuk.org
Website: www.mcsuk.org

Meat Free Mondays
'The campaign encourages people to help slow climate change, conserve precious natural resources and improve their health by having at least one plant-based day each week.'
Address: One Soho Square, London, W1D 3BQ, UK
Email: info@meatfreemondays.com
Website: www.meatfreemondays.com

Mercy for Animals
'Our mission is to construct a compassionate food system by reducing suffering and ending exploitation of animals for food.'
Address: 7908 Santa Monica Blvd, West Hollywood, CA 90046, USA
Phone: (+1) 866 632 6446
Email: info@MercyForAnimals.org
Website: www.mercyforanimals.org

National Anti-Vivisection Society
'The world's first body to challenge the use of animals in research and continues to lead the campaign today.'
Address: Millbank Tower, Millbank, London, SW1P 4QP, UK
Phone: (+44) (0) 20 7630 3340
Website: www.navs.org.uk

National Council for Voluntary Organisations
'NCVO champions the voluntary sector and volunteering because they're essential for a better society.'
Address: Society Building, 8 All Saints Street, London, N1 9RL, UK
Phone: (+44) (0) 20 7520 2414
Email: membership@ncvo.org.uk
Website: www.ncvo.org.uk

National Society for Clean Air
'Environmental Protection UK is a national charity that provides expert policy analysis and information on air quality, land quality and noise and their effects on people and communities in terms of a wide range of issues including public health, planning, transport, energy and climate.'
Address: 1 Samian Way, Stoke Gifford, Bristol, BS34 8UQ, UK
Phone: (+44) (0) 7522 840860
Email: info@environmental-protection.org.uk
Website: www.environmental-protection.org.uk

National Trail
'National Trails pass through some of the most stunning and diverse landscapes in Britain.'
Address: Walk Unlimited, 63 Delph Hill Road, Halifax, HX2 7EE, UK
Email: nationaltrails@walk.co.uk
Website: www.nationaltrail.co.uk

The National Trust
'We look after nature, beauty and history for the nation to enjoy.'
Address: P.O. Box 574, Manvers, Rotherham, S63 3FH, UK
Phone: (+44) (0) 344 800 1895
Email: enquiries@nationaltrust.org.uk
Website: www.nationaltrust.org.uk

Noise Abatement Society
'The objective of the Noise Abatement Society is to raise awareness of, and find solutions to, noise pollution and

pollutants related to solving noise issues, for example light disturbance and air pollution.'
Address: 44 Grand Parade, Brighton, East Sussex, BN2 9QA, UK
Phone: (+44) (0) 1273 823 850
Email: info@noise-abatement.org
Website: www.noiseabatementsociety.com

The Ocean Cleanup

'Our aim is to reduce 90 per cent of the floating plastic pollution by 2040, which requires global initiative.'
Address: Batavierenstraat 15, 4–7th floor, 3014 JH Rotterdam, the Netherlands
Website: www.theoceancleanup.com

OddBox

'Delicious misshapen and surplus produce delivered straight to your door.'
Address: Flat 5, 155 Endlesham Road, London, SW12 8JN, UK
Phone: (+44) (0) 20 3318 4987
Email: hello@oddbox.co.uk
Website: www.oddbox.co.uk

OLIO

'OLIO connects neighbours with each other and with local businesses so surplus food can be shared, not thrown away.'
Address: OLIO Exchange Limited, 167 Nelson Road, London, N8 9RR, UK
Email: hello@olioex.com
Website: www.olioex.com

Orangutan Care Centre

'OFI is profoundly committed to the welfare of all orangutans, whether captive, ex-captive, or wild.'
Address: 824 Wellesley Ave, Los Angeles, CA 90049, USA
Phone: (+1) 310 820 4906
Email: info@orangutan.org
Website: www.orangutan.org

Orangutan Foundation

'Orangutan Foundation has a unique and diverse approach to orangutan conservation. We support all orangutan species and our work is carried out by a team of dedicated Indonesian staff.'
Address: 7 Kent Terrace, Marylebone, London, NW1 4RP, UK
Phone: (+44) (0) 20 7724 2912
Email: info@orangutan.org.uk
Website: www.orangutan.org.uk

Organic Farmers & Growers CIC

'OF&G certify more than half of UK organic land & provide support, information and licensing to Britain's top organic food businesses.'
Address: Old Estate Yard, Albrighton, Shrewsbury, Shropshire, SY4 3AG, UK
Phone: (+44) (0) 1939 291 800
Email: info@ofgorganic.org
Website: www.ofgorganic.org

Organic Food Federation

'One of the UK's leading certification bodies operating nationally in all areas of organics.'
Address: 31 Turbine Way, EcoTech Business Park, Swaffham, Norfolk, PE37 7XD, UK
Phone: (+44) (0) 1760 720 444
Email: info@orgfoodfed.com
Website: www.orgfoodfed.com

OXFAM – BehindtheBrands

'Oxfam's Behind the Brands campaign aims to provide people who buy and enjoy these products with the information they need to hold the Big 10 to account for what happens in their supply chains.'
Address: Oxfam International, Oxfam House, John Smith Drive, Oxford, OX4 2JY, UK
Phone: (+44) (0) 1865 780 100
Website: www.behindthebrands.org

The People's Trust for Endangered Species

'We are governed by a board of volunteer trustees and our team of staff are passionate about protecting endangered species and places.'

Address: 3 Cloisters House, 8 Battersea Park Road, London, SW8 4BG, UK
Phone: (+44) (0)20 7498 4533
Email: enquiries@ptes.org
Website: www.ptes.org

Permaculture Association

'Whether you are involved in gardening, farming, planning, building homes, economic regeneration, or business, using permaculture design empowers you to make the right choices for a sustainable future.'

Address: BCM Permaculture Association, London, WC1N 3XX, UK
Phone: (+44) (0) 113 2307461
Email: communicate@permaculture.org.uk
Website: www.permaculture.org.uk

Permaculture Magazine

'Earth care, people care, future care; this is a bestselling international green/environmental magazine.'

Address: The Sustainability Centre, Droxford Road, East Meon, Hampshire, GU32 1HR, UK
Phone: (+44) (0)173 082 3311
Email: enquiries@permaculture.co.uk
Website: www.permaculture.co.uk

PETA

'The People for the Ethical Treatment of Animals (PETA) Foundation is a UK-based charity dedicated to establishing and protecting the rights of all animals.'

Address: Cannon Place, 78 Cannon Street, London, EC4N 6AF, UK
Phone: (+44) (0) 20 7837 6327
Website: www.peta.org.uk

Pod

'Pod is a UK based organisation which accepts volunteers from all over the world. We have many volunteers who join Pod projects from outside of the UK.'

Address: 3 Crescent Terrace, Cheltenham, GL50 3PE, UK
Phone: (+44) (0) 1242 241 181
Email: info@podvolunteer.org
Website: www.podvolunteer.org

Projects Abroad UK

'Projects Abroad is a social enterprise which means we employ a business model to tackle the world's social and environmental challenges.'

Address: Telecom House, 125–135 Preston Road, Brighton, BN1 6AF, UK
Phone: (+44) (0) 1273 007 230
Email: info@projects-abroad.co.uk
Website: www.projects-abroad.co.uk

Rainforest Action Network

'Rainforest Action Network preserves forests, protects the climate and upholds human rights by challenging corporate power and systemic injustice through frontline partnerships and strategic campaigns.'

Address: 425 Bush Street, Suite 300, San Francisco, CA 94108, USA
Phone: (+1) 415 398 4404
Email: answers@ran.org
Website: www.ran.org

Rainforest Alliance

'We are building an alliance to protect forests, improve the livelihoods of farmers and forest communities, promote their human rights, and help them mitigate and adapt to the climate crisis.'

Address: Warnford Court, 29 Throgmorton Street, London, EC2N 2AT, UK
Phone: +44 (0) 20 7947 4909
Website: www.rainforest-alliance.org

Recycle Now

'Recycle Now is the national recycling campaign for England, supported and funded by Government, managed by WRAP and used locally by over 90 per cent of English authorities.'
Address: Second Floor, Blenheim Court, 19 George Street, Banbury, Oxfordshire, OX16 5BH, UK
Phone: (+44) (0) 1295 819900
Website: www.recyclenow.com

Refill UK

'Refill offers a free and easy solution for preventing plastic pollution from single-use bottles at source.'
Address: Unit D, Albion Dockside Building, Hanover Place, Bristol, BS1 6UT, UK
Phone: (+44) (0)117 422 4522
Website: www.refill.org.uk

Refugee Council

'We are the only organisation providing a national service in support of refugee children and young people who arrive in the UK alone.'
Address: P.O. Box 68614, London, E15 1NS, UK
Phone: (+44) (0) 20 7346 6700
Email: info@refugeecouncil.org.uk
Website: www.refugeecouncil.org.uk

Responsible Mica Initiative

'Establish a responsible and sustainable mica supply chain in India that is free of child labor by 2022.'
Address: Espace Jacques Louis Lions, 57 avenue Pierre Sémard, 06 130 Grasse, France
Email: contact@responsible-mica-initiative.com
Website: www.responsible-mica-initiative.com

The Rivers Trust

'Our vision is wild, healthy, natural rivers valued by all.'
Address: Rain-Charm House, Kyl Cober Parc, Stoke Climsland, Callington, Cornwall, PL17 8PH, UK

Phone: (+44) (0) 1579 372 142
Email: info@theriverstrust.org
Website: www.theriverstrust.org

River Thames Society
'To conserve the river for future generations and represent all river users.'
Address: 23a Cuxham Road, Watlington, Oxfordshire, OX49 5JW, UK
Phone: (+44) (0) 1491 612456
Email: admin@riverthamessociety.org.uk
Website: www.riverthamessociety.org.uk

Road Peace
'We provide information and support services to people bereaved or seriously injured in road crashes and engage in evidence based policy and campaigning work to fight for justice for victims and reduce road danger.'
Address: 245A Coldharbour Lane, Brixton, London, SW9 8RR, UK
Phone: (+44) (0) 20 7733 1603
Email: info@roadpeace.org
Website: www.roadpeace.org

The Royal Society for the Protection of Birds (RSPB)
'Together we can give nature the home it needs.'
Address: The Lodge, Potton Road, Sandy, SG19 2DL, UK
Phone: (+44) (0) 1767 680551
Email: membership@rspb.org.uk
Website: www.rspb.org.uk

RSPCA
'Through investigations and prosecutions we stand up to those who deliberately harm animals to send out a clear message – we will not tolerate animal abuse.'
Address: Wilberforce Way, Southwater, Horsham, West Sussex, RH13 9RS, UK
Phone: (+44) (0) 300 123 0346

Email: supportercare@rspca.org.uk
Website: www.rspca.org.uk

Scottish Wildlife Trust
The Trust successfully champions the cause of wildlife through policy and campaigning work, demonstrates best practice through practical conservation and innovative partnerships, and inspires people to take positive action through its education and engagement activities.'
Address: Harbourside House, 110 Commercial St, Edinburgh, EH6 6NF, UK
Phone: (+44) (0) 131 312 7765
Email: enquiries@scottishwildlifetrust.org.uk
Website: www.scottishwildlifetrust.org.uk

Sea Life Trust
'Our vision is of a world where our oceans are healthy, properly protected and full of diverse life.'
Address: Link House, 25 West Street, Poole, Dorset, BH15 1LD, UK
Email: help@sealifetrust.com
Website: www.sealifetrust.org

Sea Turtle Conservancy
'It is the mission of Sea Turtle Conservancy to ensure the survival of sea turtles through research, education, training, advocacy and protection of the natural habitats upon which they depend.'
Address: 4581 NW 6th St, Suite A. Gainesville, FL 32609, USA
Phone: (+1) 352 373 6441
Email: stc@conserveturtles.org
Website: www.conserveturtles.org

Shakti Sings Choir
'Honouring the Earth with Song. Meets once a month in Staines, a family centred choir that sings for the joy of the earth. Raises money for charity. Includes Sing to the Land, ritual events at sacred places throughout the UK.'

Address: c/o St Peter's Community Centre, Laleham Road, Staines-upon-Thames, Staines, TW18 2DX, UK
Facebook: Shakti Sings Choir
Email: admin@shaktisings.org
Website: www.shaktisings.org

Soil Association

'The Soil Association is the UK's leading membership charity campaigning for healthy, humane and sustainable food, farming and land use.'
Address: [England & Wales]: Spear House, 51 Victoria Street, Bristol, BS1 6AD, UK
Phone: (+44) (0) 300 330 0100
Address: [Soil Association Scotland]: 20 Potterrow, Edinburgh, EH8 9BL, UK
Phone: (+44) (0) 131 370 8150
Email: feedback@soilassociation.org
Website: www.soilassociation.org

South West Coast Path

'Protecting and improving the South West Coast Path as one of the world's greatest trails, now and for future generations.'
Address: Unit 11, Residence 2, Royal William Yard, Plymouth, PL1 3RP, UK
Phone: (+44) (0) 175 289 6237
Email: hello@southwestcoastpath.org.uk
Website: www.southwestcoastpath.org.uk

Stop Ecocide Earth

'Working towards the establishment of a crime of Ecocide at the International Criminal Court (ICC), to sit alongside War Crimes, Genocide, Crimes Against Humanity and – more recently – Crimes of Aggression.'
Address: The Base, Willow Court, Beeches Green, Stroud, Gloucestershire, GL5 4BJ, UK
Email: allies@stopecocide.earth
Website: www.stopecocide.earth

Sustain
'An advocacy group promoting healthy practices in agriculture.'
Address: The Green House, 244–254 Cambridge Heath Road, London, E2 9DA, UK
Phone: (+44) (0) 20 3559 6777
Email: sustain@sustainweb.org
Website: www.sustainweb.org

Sustainable Food Trust
'Our vision is for food and farming systems which nourish the health of the planet and its people.'
Address: 38 Richmond Street, Totterdown, Bristol, BS3 4TQ, UK
Phone: (+44) (0) 117 987 1467
Email: info@sustainablefoodtrust.org
Website: www.sustainablefoodtrust.org

Sustain Your Style
'Sustain Your Style promotes sustainable fashion innovations, practices and brands with the objective of encouraging the shift of the fashion industry towards a more sustainable model.'
Email: info@sustainyourstyle.org
Website: www.sustainyourstyle.org

Sustrans
'We're the charity making it easier for people to walk and cycle.'
Address: 2 Cathedral Square, College Green, Bristol, BS1 5DD, UK
Phone: (+44) (0) 117 926 8893
Email: reception@sustrans.org.uk
Website: www.sustrans.org.uk

TerraCycle
'We partner with individual collectors such as yourself, as well as major consumer product companies, retailers, manufacturers, municipalities, and small businesses across 20 different countries.'
Address: 5 Wadsworth Road, Perivale, Greenford, UB6 7JD, UK
Phone: (+44) (0) 14 6591 5018

Email: customersupport@terracycle.co.uk
Website: www.terracycle.com/en-GB

Textile Exchange
'A global non-profit that works closely with our members to drive industry transformation in preferred fibres, integrity and standards and responsible supply networks.'
Address: Narrow Quay House, Narrow Quay, Bristol, England, BS1 4QA, UK
Email: communications@textileexchange.org
Website: www.textileexchange.org

This is Rubbish
'This is Rubbish works to reduce industry and supply chain food waste in the UK and beyond, using policy campaigns, education, the arts and public events.'
Email: info@thisisrubbish.org.uk
Website: www.thisisrubbish.org.uk

TooGoodToGo
'Save delicious food and fight food waste.'
Address: 11 Curtain Road, London, EC2A 3LT, UK
Phone: (+44) (0) 20 8089 8500
Email: info@toogoodtogo.co.uk
Website: www.toogoodtogo.co.uk

Transition Network
'Transition is a movement of communities coming together to reimagine and rebuild our world.'
Address: 43 Fore Street, Totnes, TQ9 5HN, UK
Phone: +44 (0) 1803 865 669

The Tree Council
'Working in partnership with communities, organisations and government to make trees matter to everyone.'
Address: 4 Dock Offices, Surrey Quays Road, London, SE16 2XU, UK
Phone: (+44) (0) 20 7407 9992

Email: info@treecouncil.org.uk
Website: www.treecouncil.org.uk

Trees for the Future
'We've always believed in the power of trees.'
Address: 1400 Spring Street, Suite 150, Silver Spring, MD 20910, USA
Phone: (+1) 301 565 0630
Email: info@trees.org
Website: www.trees.org

TreeSisters
'We focus on the creation of resources and experiences that empower women to step into their feminine Nature-based leadership, in support of humanity's identity shift from a consumer species to a restorer species.'
Address: Fifth Floor, Mariner House, 62 Prince Street, Bristol, BS1 4QD, UK
Email: support@treesisters.org
Website: www.treesisters.org

The Trussell Trust
'We support a nationwide network of food banks and together we provide emergency food and support to people locked in poverty, and campaign for change to end the need for food banks in the UK.'
Address: Unit 9, Ashfield Trading Estate, Ashfield Road, Salisbury, SP2 7HL, UK
Phone: (+44) (0) 20 7834 1731
Email: enquiries@trusselltrust.org
Website: www.trusselltrust.org

UK Cycle Campaign Network
'The Cycle Campaign Network is the UK national federation of cycle campaign groups, supporting cycling locally, regionally, nationally and in Europe.'
Address: Bradley Chapman, 86 Friar Street, Cleeton St Mary, Kidderminster, DY14 7RZ, UK

Email: info@cyclenetwork.org.uk
Website: www.cyclenetwork.org.uk

UK Student Climate Network
'A group of mostly under 18s taking to the streets to protest the government's lack of attention on the Climate Crisis.'
Address: 8 Delancey Passage, London, NW1 7NN, UK
Email: ukstudentsclimatenetwork@gmail.com
Website: www.ukscn.org

UNHCR – The UN Refugee Agency
'A global organization dedicated to saving lives, protecting rights and building a better future for refugees, forcibly displaced communities and stateless people.'
Address: 10 Furnival Street, Holborn, London,
EC4A 1AB, UK
Phone: (+44) (0) 20 3761 9500
Email: donors@unhcr.org
Website: www.unhcr.org

United Nations footprint calculator
'Offsetting is a climate action that enables individuals and organizations to compensate for the emissions they cannot avoid, by supporting worthy projects that reduce emissions somewhere else.'
Address: Platz der Vereinten Nationen 1, 53113 Bonn, Germany
Email: goclimateneutralnow@unfccc.int
Website: https://offset.climateneutralnow.org/footprintcalc

The Vegan Society
'One World. Many Lives. Our Choice.'
Address: Donald Watson House, 34–35 Ludgate Hill,
Birmingham, B3 1EH, UK
Phone: (+44) (0) 121 523 1730
Email: info@vegansociety.com
Website: www.vegansociety.com

The Vegetarian Society
'The Vegetarian Society is the place to go for everything you want to know about the world of vegetarian food.'
Address: Parkdale, Dunham Road, Altrincham, WA14 4QG, UK
Phone: (+44) (0) 161 925 2000
Email: hello@vegsoc.org
Website: www.vegsoc.org

Viva!
'The most powerful action you can take to end animal suffering, protect the environment and improve your health is to go vegan. Support us and help Viva! campaign for a better world.'
Address: 8 York Court, Wilder Street, Bristol, BS2 8QH, UK
Phone: (+44) (0) 117 944 1000
Email: info@viva.org.uk
Website: www.viva.org.uk

Waste Aid
'WasteAid shares waste management and recycling skills in the world's poorest places.'
Address: 8 Zealds House, 39 Church Street, Wye, Kent, TN25 5BL, UK
Phone: (+44) (0) 12 3387 7273
Email: info@wasteaid.org
Website: www.wasteaid.org

WaterAid
'Together, we can make the basic human rights of clean water, decent toilets and good hygiene normal for everyone, everywhere.'
Address: 47–49 Durham Street, London, SE11 5JD, UK
Phone: (+44) (0) 20 7793 4594
Email: supportercare@wateraid.org
Website: www.wateraid.org

Water UK
'The Water UK team engages with companies and regulators to ensure customers receive high quality tap water at a reasonable price and that our environment is protected and improved.'

Address: 3rd Floor, 36 Broadway, Westminster, London, SW1H 0BH, UK
Phone: (+44) (0) 20 7344 1844
Email: webadmin@water.org.uk
Website: www.water.org.uk

We Own It

'We Own It campaigns against privatisation and for 21st century public ownership. We believe public services belong to all of us – from the NHS to schools, water to energy, rail to Royal Mail, care work to council services.'
Address: The Old Music Hall, 106–108 Cowley Road, Oxford, OX4 1JE, UK
Phone: (+44) (0) 7923 271 431
Email: info@weownit.org.uk
Website: www.weownit.org.uk

Whale and Dolphin Conservation Society

'WDC is the leading charity dedicated to the protection of whales and dolphins.'
Address: Brookfield House, 38 St Paul Street, Chippenham, Wiltshire, SN15 1LJ, UK
Phone: (+44) (0) 1249 449500
Email: info@whales.org
Website: https://uk.whales.org

Which?

'Our founders wanted to make things better for consumers, and to raise standards across the board.'
Address: 2 Marylebone Road, London, NW1 4DF, UK
Phone: (+44) (0) 20 7770 7000
Email: support@which.co.uk
Website: www.which.co.uk

The Wildlife Trusts

'We're helping to make life better – for wildlife, for people and for future generations.'
Address: The Kiln, Mather Road, Newark, NG24 1WT, UK

Phone: (+44) (0) 1636 677711
Email: enquiry@wildlifetrusts.org
Website: www.wildlifetrusts.org

Women's Environmental Network
'Women's Environmental Network acts to achieve equality, justice and joy at the point where gender and environmental issues meet.'
Address: 20 Club Row, Shoreditch, London, E2 7EY, UK
Phone: (+44) (0) 20 7481 9004
Email: info@wen.org.uk
Website: www.wen.org.uk

Women's Health Concern
'We provide a confidential, independent service to advise, reassure and educate women of all ages about their gynaecological and sexual health, wellbeing and lifestyle concerns.'
Address: Spracklen House, East Wing, Dukes Place, Marlow, Buckinghamshire, SL7 2QH, UK
Phone: (+44) (0) 1628 890 199
Email: admin@bms-whc.org.uk
Website: www.womens-health-concern.org

Woodcraft Folk
'Our name refers to the skill of living close to nature, in the open air – we don't make a habit of hugging trees though occasionally we have been known to craft things out of wood and we do encourage all our members to come and camp with us.'
Address: 9/10, 83 Crampton Street, Walworth, London, SE17 3BQ, UK
Phone: (+44) (0) 20 7703 4173
Email: info@woodcraft.org.uk
Website: www.woodcraft.org.uk

Woodland Trust
'Woods and trees are essential. For people. For wildlife. For life.'
Address: Kempton Way, Grantham, Lincolnshire, NG31 6LL, UK

Phone: (+44) (0) 330 333 3300
Email: enquiries@woodlandtrust.org.uk
Website: www.woodlandtrust.org.uk

Work & Play Scrapstore
'Artful Reuse: through creativity, learning, design, & play.'
Address: Hazelhurst Estate, 13 Blackshaw Road, London, SW17 0DA, UK
Phone: (+44) (0) 20 8682 4216
Email: info@workandplayscrapstore.org.uk
Website: www.workandplayscrapstore.org.uk

World Economic Forum (DAVOS)
'The Forum engages the foremost political, business, cultural and other leaders of society to shape global, regional and industry agendas.'
Website: www.weforum.org

World Health Organization
'Working with 194 Member States, across six regions, and from more than 150 offices, WHO staff are united in a shared commitment to achieve better health for everyone, everywhere.'
Website: www.who.int

World Meat Free Week
'Try just one meal #WorldMeatFreeWeek.'
Address: Station House, Stamford New Road, Altrincham, WA14 1EP, UK
Phone: (+44) (0) 20 3755 6644
Email: info@worldmeatfreeweek.com
Website: www.worldmeatfreeweek.com

World Wide Fund for Nature (WWF)
'The World Wide Fund for Nature is an international non-governmental organization founded in 1961, working in the field of wilderness preservation, and the reduction of human impact on the environment.'
Address: Rufford House, Brewery Road, Woking, GU21 4LL, UK

Phone: (+44) (0) 1483 426444
Email: support@wwfhelp.zendesk.com
Website: www.wwf.org.uk

World Wide Opportunities on Organic Farms
'WWOOF is a worldwide movement linking volunteers with organic farmers and growers to promote cultural and educational experiences based on trust and non-monetary exchange, thereby helping to build a sustainable, global community.'
Address: P.O. Box 2207, Buckingham, MK18 9BW, UK
Website: www.wwoof.org.uk

WRAP UK
'Our mission is to accelerate the move to a sustainable, resource-efficient economy by re-inventing how we design, produce and sell products, re-thinking how we use and consume products, and redefining what is possible through re-use and recycling.'
Address: Second Floor, Blenheim Court, 19 George Street, Banbury, OX16 5BH, UK
Phone: (+44) (0) 1295 819900
Website: www.wrap.org.uk

XR – Extinction Rebellion
'Extinction Rebellion is an international movement that uses non-violent civil disobedience in an attempt to halt mass extinction and minimise the risk of social collapse.'
Address: The Exchange, Brick Row, Stroud, Gloucestershire, GL5 1DF, UK
Email: extinctionrebellion@risingup.org.uk
Website: www.rebellion.earth

The Young People's Trust for the Environment
'We want to give young people a real awareness of environmental problems, such as climate change, disappearing wildlife, the pollution of soil, air and water, the destruction of rainforests and wetlands, the spread of desert regions and the misuse of the oceans.'

Address: Yeovil Innovation Centre, Barracks Close, Copse Road, Yeovil, BA22 8RN, UK
Phone: (+44) (0) 1935 315025
Email: info@ypte.org.uk
Website: www.ypte.org.uk

Zoocheck

'Zoocheck endeavours to promote animal protection in specific situations and strive to bring about a new respect for all living things and the world in which they live.'

Address: 788 1/2 O'Connor Drive, Toronto, ON, M4B 2S6, Canada
Phone: (+1) 416 285 1744
Email: zoocheck@zoocheck.com
Website: www.zoocheck.com

ZQ Merino Standard

'Producing the world's leading ethical wool doesn't come easy, but our growers believe in farming sustainably for generations to come.'

Address: Level 2, 123 Victoria Street, P.O. Box 250-160, Christchurch, 8144, New Zealand
Phone: (+64) 3 335 0911
Email: info@nzmerino.co.nz
Website: www.discoverzq.com

SOURCES

18 **'This is our future ...** Greta Thunberg, 'The Disarming Case to Act on Climate Change', *TedxStockholm*, 15 December 2019

19 **'We believe that the escalating climate crisis ...** 'We Believe It's Time To Act, The Guardian's environmental pledge 2019', *Guardian*, 15 October 2019

56 **'Even in the vast and mysterious reaches of the sea ...** Brian Payton, 'Rachel Carson (1907–1964)', *NASA: Earth Observatory*, 13 November 2002

62 **'It also means that tobacco can no longer ...** 'Tobacco and its environmental impact: an overview', *World Health Organization*, 2017. Licence: CC BY-NC-SA 3.0 IGO

63 **'Tossing a cigarette butt on the ground ...** 'Tobacco and its environmental impact: an overview', *World Health Organization*, 2017. Licence: CC BY-NC-SA 3.0 IGO

69 **'thirty years of pep-talking ...** Greta Thunberg, 'The Disarming Case to Act on Climate Change', *TedxStockholm*, 15 December 2019

85 **'Consumers may think antibacterial washes ...** Janet Woodcock, M.D., FDA Center for Drug Evaluation and Research (CDER), 'FDA issues final rule on safety and effectiveness of antibacterial soaps', *US Food and Drug Administration (FDA)*, 2 September 2016

102 **'Do not be dismayed by the brokenness ...** L. R. Knost, 'InHumanity: Letters from the Trenches', *Memoirs*, 2020

104 **'We need leaders, not in love with money ...** Dr Martin Luther King, 'On Leadership: Dr. Martin Luther King, Teaching, and Alpha Phi Alpha', Elisa Villanueva Beard and Dr. Michael L. Lomax, *Teach For America*, 18 January 2016

105 **'The best motto to think about is not to waste ...** Kate Abbott, '"Just don't waste": David Attenborough's heartfelt message to next generation', *Guardian*, 19 October 2019

107 **'more masks than jellyfish in the waters ...** Francesca Giuliani-Hoffman, 'Conservationists warn Covid waste may result in "more masks than jellyfish" in the sea', Laurent Lombard, *CNN*, 24 June 2020

122 **'I'd like to tell my grandchildren that ...** Edward Felsenthal, 'TIME's Editor-in-chief on why Greta Thunberg is the Person of the Year', *TIME*, 23 December 2019

123 **'I am from Russia, where everyone can ...** Arshak Makichyan, 'Russian Youth Climate Activist Arshak Makichyan Freed from Jail', *Democracy Now*, 26 December 2019

123 **'We as kids, we're not too young ...** Portia Crowe, 'As Greta Thunberg inspires a world revolution, one young Ugandan is bringing the climate fight home', *Independent*, 28 December 2019

123 **'If we don't speak up now, how much worse ...** Autumn Peltier 'Indigenous teen only Canadian up for International Children's Peace prize', *CTV News*, 10 October 2017

123 **'We shouldn't have to tell people ...** Quannah Chasinghorse, 'Meet Quannah Chasinghorse: The 17-Year-Old Leading Climate Activism in Alaska', *The Rising*, 25 December 2019

124 ***Mni Wiconi*, meaning 'Water is Life' ...** Bobbi Jean Three Legs, 'Water is Life, Water is Sacred: Standing Rock's Bobbi Jean Three Legs Speaks Out Against Trump', *Democracy Now*, 25 January 2017

124 **'The climate crisis is not just an environmental ...** Rebecca Sabnam, 'Bangladeshi-American teen activist fighting for climate action', Samira Sadeque, *Al Jazeera News*, 3 December 2019

124 **'very significant results' by Professor Jay Mellies ...** Chris Lydgate '90, 'Bio Major Breeds Microbes That Eat Plastic. Hungry bacteria thrive on plastic water bottles, opening up the possibility of using microorganisms to fight pollution', *Reed Magazine*, 2 May 2018

125 **'children who historically don't have a voice ...** Jake Woodier, Climate Report on *BBC Newsround*, 16 February 2019

126 **'this is something that has been ...** Vanessa Nakate, 'Cropped-out climate activist Vanessa Nakate urges media to hear Africans', Jessica Murray, *Guardian*, 31 January 2020

126 **'Almost 20 million people have fled ...** Ndoni Mcunu, 'Greta Thunberg puts Africa's climate activists in media spotlight', Gaël Branchereau, *The Jakarta Post*, 3 February 2020

128 **'extensive loss, damage to or destruction ...** Polly Higgins, 'Proposed Amendment To The Rome Statute', *UK Ecocide Law*, 2010–2018

129 **media called Britain the 'dirty man of Europe' ...** Craig Bennett, Friends of the Earth, 'Outside the EU, the UK could again be the "dirty man of Europe", Devon, *The Ecologist Magazine*, 25 January 2016

129 **Laws may well be re-written to allow ...** Robert Watts, 'Cuadrilla packs up in Preston, and UK fracking bites the dust', *The Times*, 12 October 2019

131 **fracking has recently been declared 'dead' ...** Robert

Watts, 'Cuadrilla packs up in Preston, and UK fracking bites the dust', *The Times*, 12 October 2019

131 **'This is going to be my last Extinction Rebellion . . .** Judge Noble, 'Judge praises Extinction Rebellion activists after letting them walk free', Sam Courtney-Guy, *Metro*, 9 February 2020

142 **'It isn't enough just looking for . . .** Orsola de Castro, designer and cofounder of Fashion Revolution, 'Queen Of The Green Scene: Meet Ethical Fashion Pioneer Orsola De Castro', *Eluxe Magazine*, 23 January 2018

148 **'This is problematic, because the excess . . .** Jeff Holam, 'Five of the best . . . natural and organic fertilisers', *The Ecologist*, 20 May 2011

160 **'Today's business as usual is turning . . .** Greta Thunberg, 'At Davos we will tell world leaders to abandon the fossil fuel economy', World Economic Forum, 10 January 2020

166 **A Scottish scientist in the eighteenth century . . .** James E. Lovelock, 'Biodiversity', National Center for Biotechnology Information, 1988, Chapter 56: The Earth as a Living Organism, National Academies Press, Wilson E. O., Peter F. M., editors.

174 **'And yes I know that we need a system . . .** Greta Thunberg, 2019, Charlie Smith, 'Transcription of Speech to the Brilliant Minds Conference', Greta Thunberg at the Brilliant Minds Conference, Stockholm, 13 June 2019

175 **'Protest is fundamental to democracy . . .** 'Liberty Responds To Police Listing Extinction Rebellion As Extremist', *Liberty Human Rights*, Statement made in London, 10 January 2020

182 **'Trying to reverse the declines in . . .** Dr James Robinson, the Director of Conservation at the Wildfowl and Wetlands Trust, 'State of Nature Report 2019: what it means for wetlands', *Wildfowl & Wetlands Trust*, 4 October 2019

192 **'We want to facilitate synthetic fuel . . .** Stephen Beard, 'A new jet fuel offers the prospect of no-carbon, "guilt-free" flying', Minnesota Public Radio, *Market Place*, 10 October 2019

203 **'Anyone who cares about the health . . .** Marla Spivak, 'What will happen if the bees disappear?', *CNN*, 6 March 2015

218 **'There is no such thing as "away" . . .** Annie Leonard, from the documentary film *The Story of Stuff*, produced by Erica Priggen, directed by Louis Fox, released December 4, 2007 (online) and edited by Braelan Murray

219 **the area was a 'significant contamination site' . . .** Rod Minchin, 'More than 1,200 landfill sites at risk of spilling into sea on UK coastline', *Independent*, 17 February 2020

220 **In 2018 Jenny Jones from the . . .** Rob Cole, 'Incineration to outstrip recycling by 2020', *Green World*, 17 July 2018

230 **Greenpeace UK's chief scientist said . . .** Dr Doug Parr, '£3bn price rise for cost of building Hinkley Point nuclear power station', *Engineering and Technology Magazine*, 25 September 2019

279 **'The ocean is expected to contain . . .** 'The New Plastics Economy: Rethinking The Future Of Plastics', report published by The Ellen MacArthur Foundation, 2016

292 **Bolt Threads explains that 'we develop . . .** 'Bolt Technology – Meet Microsilk™', www.boltthreads.com, 2019

306 **Paul Pavol, an auto parts salesman . . .** Mike Gaworecki, 'Papua New Guinea activist receives prestigious award for protecting forests', *Mongabay, News and Inspiration from Nature's Frontline*, 18 October 2016

320 **Georgie Badiel Foundation was set up . . .** Autumn Adeigbo,'This World Water Day, Meet A Top Model Who Has Made Water Accessible To 100K People In Africa', *Forbes Magazine*, 22 March 2018

322 **'You have to listen to the science and . . .** Alexandria Villaseñor, organiser of FridaysForFuture New York and founder of Earth Uprising, '14-Year-Old Alexandria Villaseñor Has Been Striking Outside UN Headquarters For 5 Months. Here's Why', *Earth Day 2020*, 23 May 2019

328 **'there is no need for any wild animal . . .** 'Letter: Zoos are human-centric – their role in conserving species is not significant', Damian Aspinall, *Financial Times*, 11 September 2019

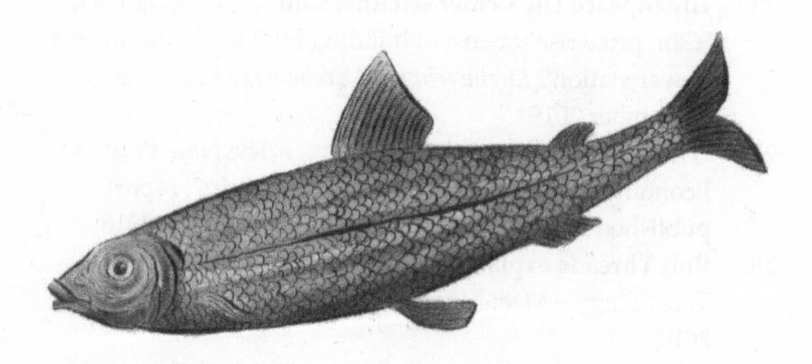

ACKNOWLEDGEMENTS

This book is a synthesis of the core environmental issues relevant today. We are grateful and humbled by the vast array of charities, groups, journalists, scientists, doctors, writers and passionate campaigners around the world whose work gives us such hope and inspiration. Their dedication and commitment to science and ecology shine through and inspire us to know more. We have acknowledged and promoted many excellent groups here who have worked hard to share their passion and information.

Thanks also to Lennie Goodings at Virago for her enthusiasm for this work and its re-emergence in a different skin from when it was first published thirty years ago, and for knowing the value of the words and message, which is ever more important. And gratitude in abundance to our good friends and eco-mentors who have read these pages, checking and tweaking until we got it right, most especially to Jean Vidler and John May and all the specialists who read selected chapters. We value all your time and effort with respect and our thanks.

Thanks, too, go to our families and friends who have supported this journey and brought us all what we need so we could produce these words for you to enjoy. Such love is precious and generous and has enabled us to share our words with you, dear reader. So, many thanks and with all our hearts, to Alexander, Michael, Alexandria, Kate and Dudley. And to Amy's family and friends, Mum, Dad, Henry, Esua, and everyone who made this possible.

Thank you especially to Stu McLellan for his observations and delicious artworks, which embellish this book with such joy, despite the species emergency they represent.

Finally, this book is being written in an unprecedented time on earth. It would be wrong of us not to extend sincere gratitude and love to those in our society who have supported, served, protected, nursed, fed and loved us throughout the most significant crisis humanity has faced in this generation. Our wish is that this is a wake-up call, and that an extraordinary new ecological respect will rise from humankind to all creatures on earth.

And if you are on that journey with us, we thank you too.